B Ingram.

Dr. H. C. King's
BOOK OF ASTRONOMY

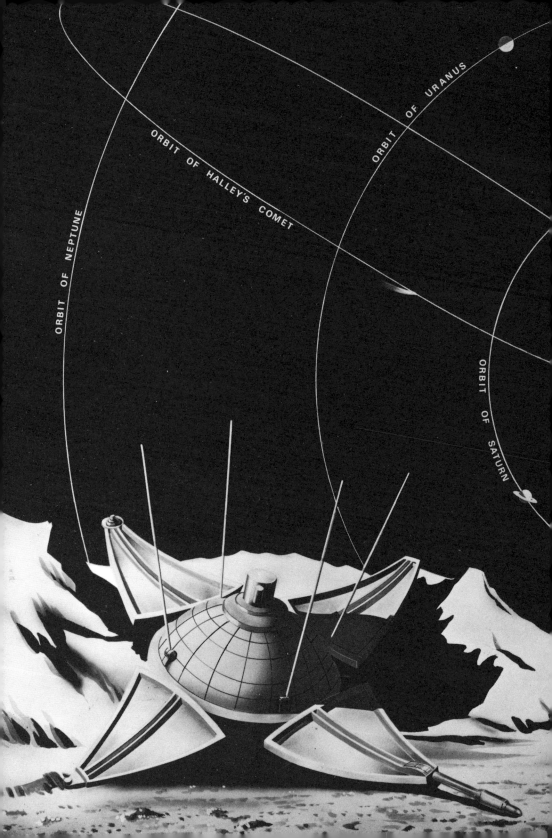

Dr. H. C. King's
BOOK OF ASTRONOMY

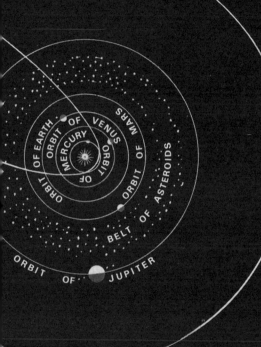

COLLINS
LONDON AND GLASGOW

FIRST PUBLISHED IN 1966

DR. HENRY C. KING, M.SC., F.R.A.S.
IS SCIENTIFIC DIRECTOR, THE LONDON PLANETARIUM

© DR. H. C. KING, 1966

ISBN 0 00 106114 3

PRINTED AND MADE IN GREAT BRITAIN
BY WILLIAM COLLINS SONS & CO. LTD.
LONDON AND GLASGOW

Contents

    Introduction · 7
1   The Shape and Size of the Earth · 9
2   Earth, a Rotating Ball · 17
3   Planet Earth · 26
4   The Telescope · 36
5   The Spectrograph · 43
6   The Moon · 52
7   Five Naked-eye Planets · 65
8   Additions to the Solar System · 79
9   Comets and Meteors · 87
10 The Sun · 95
11 A System of Stars · 108
12 Gas, Dust and Stars · 121
13 Stars and Galaxies · 133
    Appendix A · 145
    Appendix B · 146
    Glossary · 147
    Bibliography · 155
    Index · 156
    Acknowledgements · 160

Introduction

ASTRONOMY is a scientific discipline. A great deal of it is concerned with physics, much more with mathematics, and only a very small part with looking through a telescope. This perhaps explains why some people hear and read about it with a feeling of bewilderment. They had little or no science in their schooldays, and quite naturally find its language strange and difficult to follow. As more than one visitor to the London Planetarium has remarked: 'When you astronomers talk about light-years and the expanding universe I give up'. But the root of the problem is that most of us tend to 'give up' rather too quickly. We either underestimate our mental abilities or get lazy in our thinking. Which is a pity. For on that basis the modern sky with all its majesty, immensity and complexity must ever remain a closed book. We might as well live on Venus' a planet so cloudy that we would be completely shut off from the rest of the universe.

This is where young people score. Most of them take elementary physics or general science at school. They are therefore introduced to the methods and language of science at an early age and learn to accept such words as 'energy', 'spectrum', 'motion', 'velocity' and 'light-year' as a necessary part of its discipline. For them this book should be easy reading. One hopes, of course, that it will reach a far wider audience. Astronomy *is* a science but it carries no sign: 'Non-Specialists Keep Out!'

In my schooldays, over thirty years ago, astronomy was hardly ever mentioned in school. We boys got the idea that it was concerned mainly with the study of constellations, predicting eclipses of the sun and moon, and explaining how the seasons were produced. In some quarters this view still persists. In others the subject is identified with the flights of the astronauts and attempts to land men on the moon. So it is not surprising that many children (and grown-ups too) think that astronomy is 'all about' constellations, space, rockets, the moon and the universe. In fact, a boy recently asked me to send him 'information about the universe and, if possible, a photograph of Colonel Glenn'. His teacher had apparently given his class a week-end project entitled: 'Space'.

This book tries to do at least three things. First, to give the subject an historical background. Second, to compare and contrast the old picture of a fixed earth and eternal sky with the modern one of a moving earth and evolving sky. Third, to explain in simple and concise terms how astronomers are able to discover so much about bodies so far away. It is certainly no textbook nor does it pretend to cover each and every aspect of the subject. But it does deal with basic things, and these readily capture the interest and imagination of seriously-minded young people. If it does this, perhaps to the extent of encouraging its readers to make a deeper study of astronomy, my efforts will be most amply rewarded.

November, 1965.   H. C. KING

CHAPTER 1

The Shape and Size of the Earth

FOUR thousand years or so ago civilisation was at its highest in the Near and Middle East. In countries like Mesopotamia (modern Iraq) and Egypt, a visitor would have seen great temples dedicated to the worship of sky-gods like the sun and moon. Nothing was then known of astronomy as a science. True, the movements of the sun, moon and stars marked the passage of time and gave man the units of the year, month and day, but the simple reckoning of these periods could hardly be called a science. Early peoples also recognised the five naked-eye planets Mercury, Venus, Mars, Jupiter and Saturn. These, they thought, were very special stars for they wandered about among the others as if they were alive. The sun and moon also moved in relation to the stars so they too were considered to be living things. They lived up in the sky and were therefore gods or goddesses. As for the earth, this was thought to be flat and fixed. Covered over by the sky it formed, so to speak, the base of the entire universe—a universe at once small, completely shut in, ruled by gods, and centred on man.

Forget, for a moment, that you already know that the earth is round like a ball. How would you, using just your eyes and without travelling far, prove that it is not flat? The countryside looks flat, or almost so if you ignore any local hills and valleys. If somehow, you discover that it is slightly curved this still does not prove that the earth as a whole has a curved surface. It may be convex or curved outwards in one part, and concave or curved inwards in another. The ground seems solid and deeply rooted, unless, of course, you happen to live in an earthquake area. And the sky? This appears to be no more than a curved ceiling which meets the ground at or beyond the distant skyline.

The first people to break with the old idea of a flat earth were the Ancient Greeks. A number of them pointed out that things were not necessarily what they appeared to be. The earth, they suggested, might be round like a ball but so big that, to a man standing on its surface, it

Opposite The Pleiades, a conspicuous cluster of hot, blue stars wreathed in nebulosity.

appeared to be flat. In the sixth century B.C. mathematicians like Pythagoras supported this idea since, in their opinion, a sphere was the most perfect shape. They therefore considered it to be a proper shape for so worthy an object as the earth.

The new view also received support from observation, and in the fourth century B.C., when Aristotle wrote a book about the heavens, it was widely accepted not as an idea but as a fact. During a total eclipse of the moon, Aristotle wrote, the earth's shadow falls on the moon. The edge of the shadow is seen to be curved, hence the edge of the earth must also be curved. Further, stars like the North Pole Star and others in the northern part of the sky get gradually higher as we travel southwards. Conversely, they get gradually lower as we travel northwards. This, Aristotle claimed, showed quite clearly that the earth was round. For if it were flat the starry sky would look the same no matter how far north or south we travelled. A third bit of evidence, known to the Ancient Greeks but overlooked by Aristotle, was given by the appearance of distant objects at sea. As a ship approached land the hills and cliffs appeared to rise out of the sea. Whereas if the earth were flat they would be visible over a much greater distance. Clearly, then, the earth's surface was curved in and around the Mediterranean area. This being so, it was reasonable to assume that it was curved in a similar way all the way round.

For places well north of the equator the height of the midday sun is greatest at the summer solstice (June 22) and lowest at the winter solstice (December 22). The Ancient Greeks discovered that at, say, the summer solstice, the sun at midday seen from the Greek mainland appeared lower down than when seen from Egypt. This observation was to be expected in view of the earth's curvature, but to Eratosthenes, a mathematician at Alexandria in the second century B.C., it meant much more. He knew that at midday at the summer solstice, vertical pillars at Syene (modern Aswân) cast no shadows. Yet at the same time at Alexandria the vertical pillar of a sundial formed a definite shadow. By measuring the length of the shadow, and knowing the height of the pillar, he was able to calculate the sun's angular distance from the *zenith*, or point directly overhead. At Syene, then, with the sun directly overhead, its zenith distance was zero. But at Alexandria its zenith distance was $7°·2$, or one fiftieth of $360°$. He was certain that Alexandria lay due north of Syene. He also knew that professional pacers had put the distance between the

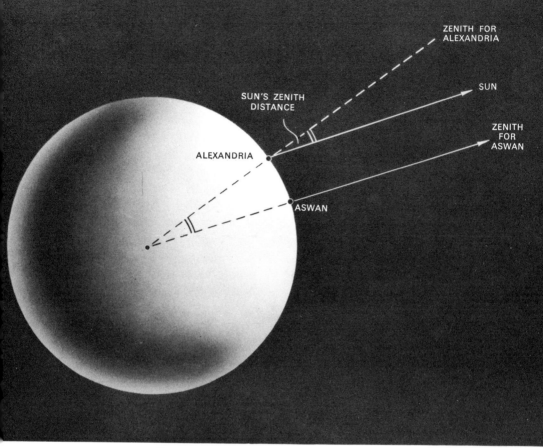

Eratosthenes' method of measuring the size of the earth.

two places at 5,000 stadia. So, he argued, if a journey of 5,000 stadia due northwards caused the sun to move through one fiftieth of 360°, one of 50 × 5,000 stadia would cause it to move through 360°, or a complete circle. In other words, and on the assumption that the sun was very far away, the circumference of the earth was 250,000 stadia. This value divided by π or 3·172 gave the earth's diameter.

The length of a stadium is not known for sure but it was most likely 517 feet. On this basis the earth's polar diameter according to Eratosthenes was 7,800 miles, or only 100 miles short of the modern value of 7,900 miles. This method is a perfectly sound one but astronomers would not think of using it today. If Eratosthenes had repeated the experiment he would most likely have got a completely different result. He could easily have got 6° or 8° instead of 7°·2 for the sun's zenith distance, in which case the earth's diameter would have come out at 9,360 and 6,920 miles respectively.

Al-Khwarizmi, a Muslim mathematician of the ninth century A.D., made the next serious attempt to measure the earth. He decided to find the length of one degree of the meridian. Once this was known he had only to multiply it by 360 to get the earth's circumference. This he did by assembling some observers on a large plain and sending them first to the north and then to the south of a base camp. In each case they measured the distance they had to travel to make the North Pole Star move one degree away from its original position. The average of the two distances, 113 miles, led to the excessive value of 13,000 miles for the earth's diameter.

The trouble with a method of this kind is that it is extremely difficult, even with modern instruments, to decide when a star has changed its zenith distance by precisely one degree. Al-Khwarizmi would have done better if, like Eratosthenes, he had aimed at a much larger change in angle. The instruments he used were probably based on a measuring device (introduced in the second century A.D. by Ptolemy) known as a *quadrant*. In this a quadrant or one quarter of a circle had its curved edge or 'limb' divided into degrees, minutes and parts of a minute of angle. It was then supported in an upright position and adjusted so that a plumb-line crossed the 90-degree mark. The observer looked along a moveable arm or radius pivoted to the quadrant's centre of curvature. Pins at the ends of the arm enabled him to 'sight' a particular star, and by so doing, to read off its zenith distance.

If you ask your friends the question: 'Who discovered that the earth is round?' you will probably be told: 'Christopher Columbus'. Yet Columbus knew that the earth was round and had a good idea of its size long before he ventured to sea. He made his great voyage westward in 1492 in the firm belief that he would eventually reach the East Indies, a belief based on the idea that the earth was round. He greatly underestimated the distance involved, took over two months on the journey and was more than relieved when he sighted land. He thought he had discovered the East Indies but the island he landed on was one of the Bahama group, now known as Watling Island. He made three voyages westward. On the second he explored the West Indies. On the third he reached a great land to the south of the Indies but never lived to know that it was the continent of South America.

THE SHAPE AND SIZE OF THE EARTH

Although mathematicians and geographers declared that the earth was a sphere, ordinary folk were never completely convinced of this until Ferdinand Magellan sailed all the way round it. He left Spain on September 20, 1519, with five ships, but only one, the little *Victoria*, survived the journey. Three years later nineteen famine-stricken survivors staggered ashore at Seville to tell of the tragic death of their commander in the Philippines and the first circumnavigation of the world.

The modern period of the study of the size and shape of the earth began early in the seventeenth century. Using methods similar to those of Eratosthenes and Al-Khwarizmi, Willebrord Snell in 1617 found that one degree of the meridian contained $66\frac{2}{3}$ miles. In 1635 Richard Norwood measured the distance from London to York with a chain of known length, and from observations of the sun's zenith distance at midday, found that one degree contained $69\frac{1}{2}$ miles.

The best result was obtained in 1671 by Jean Picard of Paris. He used a quadrant fitted with a telescope instead of with plain sights. He could therefore point it more accurately at a given star. In much the same way a marksman today, using a gun fitted with telescopic sights, can aim his gun more precisely at a distant object. Picard also introduced a method known as *triangulation*. In this the distance between two places in open, flat country is carefully measured by means of a chain of standard length, or nowadays with an electronic gadget. This distance is known as a *base line*. A theodolite is then set up at each end of the base line and used to measure the direction of a prominent landmark such as a church spire or specially erected pillar. By doing so the surveyor obtains one side and two angles of the triangle formed by the base line and the distant object. The other two sides are then calculated by trigonometry and one of them is used as a base line for building up another triangle. In this way triangle is built on triangle until a whole network of them covers the region under survey. The distance between two places in different latitudes can therefore be found and compared with that given by astronomical observation. The value of all this in map-making is obvious. In Great Britain alone the main features on a coastline can now be mapped with an accuracy of a few inches. Three centuries or so ago their mapped positions were sometimes several miles in error.

Picard's method made possible a more accurate map of France. His result, 69·1 miles for one degree of the meridian, gave Isaac Newton, then working on his theory of gravitation, a reliable value for the size of the earth. It also meant that astronomers could form better estimates of the distances of the moon and planets, for as we shall see later, these estimates depended on one known distance—the radius of the earth.

In 1671 the French astronomer Jean Richer went on a scientific expedition to Cayenne in French Guiana. While he was there he noticed

Method of triangulation.

that a pendulum clock lost time, although the same clock at Paris had given correct time. He realised that the fault lay not with the clock but rather with the earth. The pendulum, he suggested, swung more slowly at Cayenne than at Paris because the pull or force of gravity was less near the equator than in higher latitudes. This in turn indicated that the earth was not a true sphere. What, then, *was* its shape? Newton, applying his theory of gravitation, found that the earth had a slight bulge in the region of its equator. More precisely, its average shape was an oblate spheroid, or the figure obtained when an ellipse is rotated on its shorter or minor axis. If this were so, a degree of latitude would be shortest (68·17 miles) at the equator and longest (69·40 miles) at the poles. Con-

Shape of the earth.

firmation came in 1738 from the French astronomer Pierre Maupertuis. By then French expeditions had measured the length of a degree of latitude in places as far apart as Peru and Lapland. Maupertuis, with a number of different values to work with, found that they all closely fitted just the shape Newton had predicted.

Once the average form or shape of the earth had been established it was left to astronomers, surveyors and navigators to find the actual shape or *geoid* as it is called. Differences in latitude continued to be found by the old method of observing differences in the zenith distances of suitable stars. Differences in longitude presented no problem after 1761 when a special marine clock or chronometer made by John Harrison successfully completed its first big trial run. The clock kept such good time that after a six weeks' sea voyage to Jamaica, its error was not more than five seconds, corresponding to an error in longitude of just over one minute of angle.

Scientists worked hard to find the exact amount of the earth's equatorial bulge, the curvature of the surface at different places, and the distances of those places from the earth's centre. This study, known as *geodesy*, involved measuring distances in both latitude and longitude, determining the relative strengths of the force of gravity at different places, and undertaking great surveys based on the method of triangulation.

At its meeting in 1924 at Madrid, the International Astronomical Union adopted a number of 'constants' for the size and shape of the earth. Three of them were:

$$\text{Equatorial radius} \quad a = 3963\cdot35 \text{ miles}$$
$$\text{Polar radius} \quad b = 3950\cdot01 \text{ miles}$$
$$\text{Polar flattening} \quad (a-b)/a = \tfrac{1}{297}$$

Although astronomers and others still use these values they do not necessarily regard them as the best possible. In recent years, for instance, they have been improved upon by studying the paths or orbits of artificial earth satellites. When a satellite orbit is tilted in relation to the earth's equator the equatorial bulge tends to pull the satellite off course. In consequence the orbit precesses, or moves round the earth in longitude. By measuring the rate at which this happens, scientists have calculated that the amount of the earth's polar flattening is about $\tfrac{1}{298\cdot4}$. Satellite orbits also show other irregularities which reveal that the earth is very slightly pear-shaped, with the 'bump' or bulge in the southern hemisphere. These latest findings have an important bearing on theories about the earth's interior and the way it has been formed. They show, for example, that the earth's mantle (the region between its liquid core and surface crust) is remarkably rigid. Otherwise it could not possibly hold up its bulges against the pull of gravity.

CHAPTER 2

Earth, a Rotating Ball

ARISTOTLE and other Greek thinkers tried to discover the nature and structure of the whole universe. To do this they first decided how the universe should be constructed and then applied their decisions on a most ambitious scale. Here is an example of their way of thinking. The circle, they said, has a perfect figure. Movement in a circle with unchanging or

The universe according to Aristotle. The spheres immediately surrounding the fixed central earth and its waters are those of air and fire.

constant speed is therefore perfect movement. The heavens are unchanging and therefore perfect. Hence the sun, moon, planets and stars travel with unchanging speeds in circular paths. These paths all have the earth at their centre, for a central position is the earth's right and proper place in the universe. It stays where it does because it has no wish or tendency to go either one way or another. The earth, unlike the heavens, is the scene of change. It is therefore imperfect. Movement along a straight line is imperfect movement. This is therefore the kind of movement found on the earth but never in the heavens.

We no longer accept the statement: 'unchanging circular movement is perfect movement' as a statement of fact. We could equally well claim: 'movement in an ellipse is perfect movement'. If that were true, and since the moon and planets are known to move in ellipses, their movements would be perfect. But what is the meaning of the word 'perfect' anyway? One might as well say that a thing that changes is perfect and one that does not is imperfect. A clock whose hands never move would not be considered perfect. All we are doing is just playing around with ideas and words, and this is precisely what Aristotle and most other Greek thinkers loved to do. Small wonder, then, that when men of a later age began to question whether Aristotle's statements really were statements of fact, his whole system of the universe began to collapse.

According to Aristotle the universe was made up of a series of crystal-like spherical shells or globes. They were placed one inside the other rather like the layers of an onion. At their common centre lay the fixed spherical earth, the scene of birth, growth, decay and death. Completely surrounding this was the shell or sphere of the moon. The earth's atmosphere reached right up to this sphere and formed a region of change in which occurred such fleeting things as rainbows, shooting stars and comets. Ranged in order above the moon's sphere were the shells of Mercury, Venus, the Sun, Mars, Jupiter and Saturn. Each shell had its own individual movement and also joined in the daily or 24-hour rotation of the outermost shell, the sphere of the fixed stars. The stars, sun, moon and planets occupied celestial regions, that is, regions divine and unchanging in contrast to the terrestrial region shut in by the sphere of the moon. Aristotle's universe was therefore finite, spherical and earth-centred. It was also of immense size, for compared with the great sphere of fixed stars the earth was considered to be a speck and man an even tinier speck.

In the light of modern knowledge Aristotle's model of the universe appears absurd to the point of being ridiculous. We should not, however, be over-hasty in our judgment but rather try to understand why he made it in this way. To the naked eye the heavens seem to be unchanging. Groups and patterns of stars look the same year after year and century after century. Stars like those of the Plough and Orion behave just as if they are fixed to the sky. The sun, moon and planets move about in a fairly regular manner. They look the same to one generation as to another. True, the stars wheel as one great company across the sky but this did not suggest, at least to the Greek mind, that the earth rotated. It was much more reasonable, said Ptolemy, to think of the stars as moving, for they are light, airy things whereas the earth is dull and heavy. If the earth rotates, he asked, why do we not feel or see any effects of its movement? Give it a rotation and straightway it would throw everything off its surface just as a spinning potter's wheel throws off bits of wet clay.

Not everyone saw eye-to-eye with Aristotle. In the fourth century B.C., for example, Heracleides and other disciples of Pythagoras thought that the daily rotation of the starry sky could be produced equally well by a spin of the earth on its axis. Aristarchus of Samos, who lived about 320-310 B.C., even went so far as to suggest that the sun and not the earth occupied the centre of the universe. The earth, he declared, revolved around the sun in a circular orbit once a year. But these ideas had a cool reception and were generally ignored right up to the middle of the sixteenth century. Until then the movements in the sky could be quite satisfactorily explained on the basis of a fixed central earth.

The first break with the fixed-earth viewpoint came in 1543 at the hands of the Polish astronomer Copernicus. In a book published in that year, Copernicus gave reasons for thinking of the earth as a planet. In his opinion it not only rotated on its axis but also revolved round the sun. Here we shall deal with only the first of these views, that is, with the rotation, leaving the second over for discussion in the next chapter.

Copernicus realised, as did some of the Ancient Greeks, that if the earth's surface moved in a particular direction we should expect to see distant bodies like the stars move in an opposite direction. The daily rotation of the sky, he maintained, is an obvious example of this. The rising of the sun, moon and planets in the east and their movement across the sky to set in the west is no more than an apparent movement produced by the earth's spin from west to east. The stars move in a similar way, but with one important difference—those in the northern part of the sky move from west to east, or in the *same* direction as the earth's spin. But this is no great problem. It so happens that the earth's axis, carried through the north pole, points roughly to *Polaris*, the North Pole Star. Seen from mid-northern latitudes this star is fairly high up. More precisely, its altitude or angular height above the northern horizon is roughly equal to the latitude. At the equator, latitude 0°, *Polaris* is near the horizon. At the north pole, latitude 90°, it is near the zenith.

So as the earth rotates the stars are seen to travel round *Polaris* as if tied to it by invisible strings. In other words, the movement of the starry sky is not a simple east-to-west one but a turning or wheeling motion centred on *Polaris*. This is exactly what we should expect to see if the earth rotated daily on its axis and had that axis aimed at *Polaris*. Actually, and as Copernicus realised, *Polaris* is not at the *north celestial pole* or actual turning point of the northern sky. At present it is about one degree away from this point, but for the purposes of ordinary naked-eye observation the difference can be ignored.

Copernicus knew that he would have to face up to Ptolemy's argument that a spinning earth would throw things off its surface. Simple calculation shows that objects at the equator are carried round at about 1,000 miles an hour. At this speed a stone thrown straight up should come down several hundred feet westward of the place from where it was thrown. Why, then, do things thrown vertically up always come straight down? Copernicus replied that they do so because they are all carried round with the earth. Everything goes round—trees, stones, houses, people, even the air—hence we get the impression that these things are at rest. Another objection was that so rapid a motion would cause the earth to fly apart. Why then, he asked, doesn't the sky fly apart? For if the earth is fixed the sphere of stars *must* rotate. Since the sky is very much larger than the earth, points on its surface must be moving at terrific speeds. It would therefore be in a much greater danger of breaking up than the tiny spinning earth.

Copernicus thought that the stars were all at the same distance from us. He therefore adopted Aristotle's idea of a universe shut in by the great sphere of fixed unchanging stars. Towards the end of 1572, however, twenty-nine years after his death, a bright star appeared and quickly rose

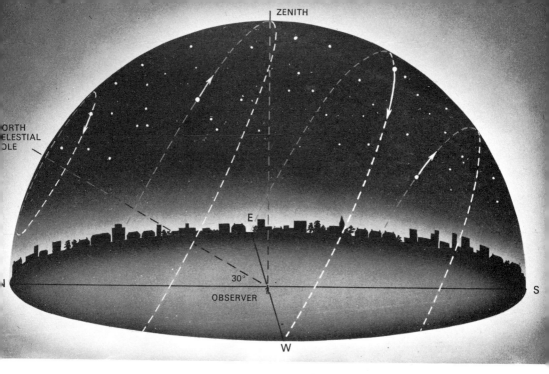

Apparent motion of the stars.

in brilliance to outshine Venus, the brightest planet. Since Aristotle had declared that the heavens were unchanging, everyone at first thought that the object was some sort of explosion in the earth's atmosphere. Astronomers tried to find how far away it was. A fairly near object would show a definite *parallax*, that is, it would be seen in one direction from one particular place and in a different direction from another. They measured its position relative to the stars from a number of widely separated places but found no change of any kind. The conclusion was obvious. The object belonged to the sphere of fixed stars. Aristotle and his followers had therefore been wrong in thinking that the stars never changed.

The star of 1572 slowly faded and in March, 1574, disappeared. Another nearly as bright appeared in 1604 and several others have been seen since. They became known as 'novae' or 'new stars' but we now know that 'temporary stars' would be a better description. When a nova appears, a study of earlier photographs usually reveals a faint star precisely in the same position. The star, for reasons not yet fully understood, undergoes a sudden and immense flare up and then gradually sinks into obscurity.

Very bright ones like those of 1572 and 1604 are called *supernovae* and owe their brilliance to an explosion which affects the whole star. Less brilliant ones like Nova Aquila 1908 and Nova Herculis 1934 blow off just their surface layers.

By the early seventeenth century it had become pretty obvious that the earth exerted a pull or attractive force on objects on and near its surface. Did material bodies possess a curious kind of 'affection' towards one another? Was this why an apple fell from a tree to the ground instead of going upwards? Did the earth have a magnetic 'virtue' and attract other bodies to it in much the same way as a magnet attracts iron filings? If so, how did the pull vary with distance? Perhaps its strength was the same at all distances? Newton called the force of attraction 'gravitation' but left open the question of its nature. What mattered most was that gravitation behaved in a definite way, and by discovering the way it behaved he laid the foundations of the modern view of the structure of the physical universe.

Objects remain on the earth, Newton explained, because they are attracted to the earth by the force of gravity. Knowing this, he was able to account for Richer's observations of the change in the time of the swing of a pendulum. He found that the force of gravity at the earth's surface is slightly reduced by an opposing effect due to the earth's rotation. This opposing effect gives bodies a tendency to fly away from the earth, hence its name *centrifugal effect* or *centrifugal force*. At Cayenne, where the speed of the earth's surface is greater than at Paris, the centrifugal effect is greater and so a pendulum swings more slowly.

For the same reason, Newton found, the waters and material of the earth itself tend to be pulled outwards at and near the equator. If the spinning earth were a true sphere the seas would all flow to the equator. There they would pile up to form an extremely deep ocean. But as we have already seen, it is the earth itself and not the seas alone which is deformed in this way. Because the earth has a bulge at its equator, the seas have an almost constant level at both the poles and the equator.

Further evidence of the earth's rotation was provided by George Hadley in 1735. In his opinion, the main or prevailing winds on a non-rotating earth would be from the north in the northern hemisphere and from the south in the southern hemisphere. Instead, those near the equator came from the north-east in the northern hemisphere (the N.E. Trades) and from the south-east in the southern hemisphere (the S.E. Trades). This, Hadley explained, was because the earth rotated from west to east. The rotation produced a deviating effect (known as the *Coriolis force*) which gave the polar or north-south and south-north winds a slightly east-to-west course.

The deviating effect was nicely demonstrated in 1851 by the French physicist Léon Foucault. He hung an iron ball on a fine wire more than 200 feet long from the dome of the Panthéon in Paris. The ball was then slightly drawn aside and allowed to swing to and fro like the bob of a pendulum. A pendulum of this kind swings for a long time and keeps to a fixed direction in space. If the earth did not rotate, this direction would remain the same relative to surrounding objects. In fact, the direction relative to these objects slowly changes, and in such a way as to show that the earth rotates from east to west.

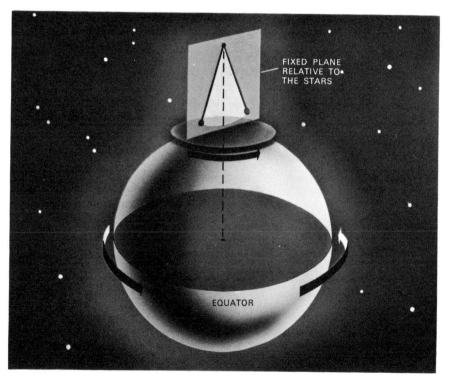

Foucault's pendulum at the North Pole.

We have also to thank Foucault for the *gyroscope*, a heavy wheel which, when made to rotate rapidly, keeps its axis of rotation in a fixed direction in space. If the axis is pointed at a star it continues to point at that star. This shows that the stars have fixed directions and that their apparent general change in direction is due solely to the earth's rotation. The earth with its slight equatorial bulge acts rather like a gyroscope. It therefore has, like a gyroscope, great stability against toppling over. This is an important property in view of the great upheavals of the past which, according to the evidence of geology, pushed up the mountains and created the depths of the oceans.

Finally, bodies dropped from high towers or down mine-shafts do not fall straight down but travel slightly eastwards. For, owing to the earth's spin, the top of a tower travels slightly faster than the ground, and the ground travels slightly faster than the bottom of a deep hole. The small difference in speed gives rise to a small but measurable deviation. It amounts to just over an inch for an object dropped from a height of 500 feet. For the same reason a stone thrown straight upwards will hit the ground at a point slightly to the east of the starting point. But in this case the deviation is too small to be noticed with the unaided eye.

Astronomers have always been interested in time and its measurement. One unit of time, the day, or period of the earth's rotation, is now known with great accuracy. Its length depends on how it is measured. The most basic way is to find how long it takes the starry sky to make one complete rotation. This is done by measuring the interval between two successive crossings or *transits* of a star across the meridian. The meridian referred to is an imaginary line drawn through the north and south points and the zenith of the place of observation. The star is observed with a *transit instrument*, or telescope mounted so that its tube can be pointed only along the meridian. In practice the method has lots of refinements and requires the use of accurate clocks of advanced design. It tells astronomers that the earth rotates once in 23 hours 56 minutes 4·099 seconds. This is known as a *stellar day*, or day based on observations of the stars.

A study of the records of ancient eclipses of the sun show that over the past 3,000 years the earth's rotation has been slowing down at the rate of 0·0014 of a second a century. The slowing down is due mainly to the friction of the ocean tides on the sea-bed, especially in shallow seas like the Bering Sea and Irish Sea. The tides in their turn are produced by the moon's gravitational pull, which can therefore be thought of as having a slowing down or braking effect on the earth's spin. So after thousands of millions of years the day could become as long as the month, or the time taken by the moon to travel once round the earth. The earth would then turn one and the same face always to the moon, just as today the moon turns the same face towards the earth.

By the same token, the earth in its early history probably rotated faster than it does now. This would have caused it to bulge more at the equator than at present. That it can change its shape even only slightly over long periods of time means that it cannot be completely stiff or rigid. Its core or innermost parts must be sufficiently liquid to allow it to take up the ellipsoidal shape. Studies of the way earthquake waves travel inside the earth indicate that this is indeed the case. The material is probably made up of iron and other heavy metals in the molten state and forms a core believed to have a radius of about 2,200 miles.

Further studies make it clear that there is no easy solution to the problem of the slowing down of the earth's spin. Calculations of the amount of friction due to the tides show that the rate of slowing down should really be 0·0023 of a second a century. Since the actual amount appears to be much smaller, namely, 0·0014 of a second a century, something else must be reducing the effect of tidal friction by about 0·001 of a second a century. The 'something else' could be a reduction in the earth's radius of spin produced by a fall in the mean level of the sea. The amount would be no more than about 6 feet in 2,000 years. But scientists are by no means agreed that this is the main cause or even a cause at all.

Observations made over the last 250 years show that certain very small changes arise from time to time in the observed and predicted positions of the sun, moon and planets. These changes or irregularities vary in amount but always occur together, that is, at the same time. At first the moon was thought to be the culprit, but the fault was finally traced to the earth. The earth makes more or less sudden changes in its period of rotation, increasing or decreasing it by as much as 0·003 of a second. Just how it does this is not known, but here again the effect could be produced by a sudden change in sea-level or in the radius of spin. At all events, it is now clear that the earth is not such a good timekeeper after all.

CHAPTER 3

Planet Earth

WE mentioned earlier that Copernicus stated that the earth not only rotated on its axis but also revolved round the sun. A few others, Aristarchus among them, had already toyed with this idea. Copernicus was the first to work out its consequences in detail. None recognised better than he that any new theory of the universe had first to meet the requirements of observation. That is, it had to account for the observed movements of the sun, moon and planets so well that their positions could be predicted with reasonable accuracy for many years ahead.

Ptolemy and other Greek astronomers before him had done just this. They followed Aristotle in placing the earth at the centre of the universe. They also accepted without question the idea that the heavenly bodies travelled with constant speeds in circular paths. On the other hand, they knew from observation that the sun, and certainly the moon and planets, did not move across the background of stars with constant speeds. To resolve the difficulty they hit on an ingenious idea. Why not let unchanging motion in *two* circles represent the observed motion of a body? Why not let the sun, say, move with constant speed on a circle whose centre moves with constant speed on another circle centred on the earth? Once this was done, and after carefully adjusting the relative sizes of the circles and the speeds on them, the arrangement worked quite well. It accounted for the irregular or unequal motion of the sun—that it travelled faster in winter than in summer. Also for the fact that it appeared slightly larger than average in winter and slightly smaller in summer—a change correctly attributed to a change in distance. A similar arrangement, but with various small additions and alterations, met the requirements of the moon and each of the five planets. In the end, of course, the whole theory became quite involved. But this was a small price to pay for a model that represented all the then-known movements in the sky.

Copernicus was bold enough to question the old view that the earth-centred system was the only possible one. Why not, he asked, make the sun

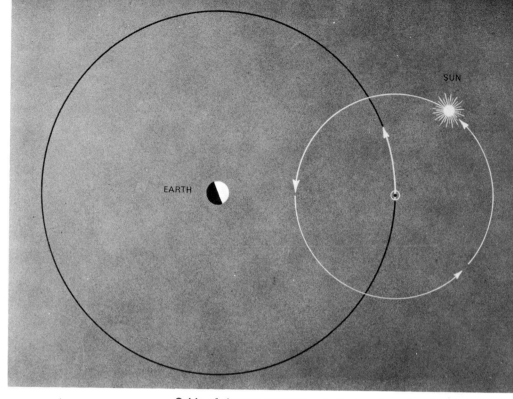

Orbit of the sun according to the Ancient Greeks.

the central body around which all the planets travel, the earth included? The sun's size and brilliance qualify it for this position. From the centre it can best control and light up its wheeling family of planets. Only the moon goes round the earth and, by so doing, ranks as an attendant or satellite of the earth. As for the stars, they are fixed to the sphere of the sky as Aristotle and others have taught. The starry sphere therefore encloses the universe and has the sun at its centre. It is of immense size compared with the solar system, that is, with the system of the sun and its six planets.

Copernicus put the planets Mercury, Venus, Earth, Mars, Jupiter and Saturn in this order of distance from the sun. He was guided, correctly, by the fact that this is the order of their periods of revolution round the sun. Mercury, the swiftest, travels once round in an average time of 88 days while Venus takes 225. Seen from the earth, these two planets never get very far from the sun, although Venus can move farther away than Mercury. They appear as 'Morning Stars' when they rise before the sun and as 'Evening Stars' when they set after the sun. Clearly, then, their orbits lie inside the earth's orbit, with that of Mercury closer to the sun than that of Venus. Mars, Jupiter and Saturn have periods of about 2, 12 and $29\frac{1}{2}$ years respectively. Seen from the earth they can wander well clear of the sun and often appear in the south long after the sun has set. Their orbits therefore lie outside the earth's orbit.

According to this scheme the slow movement of a planet against the background of stars is a combination of two movements—one of the planet itself and one due to the earth. The observed change in brightness of a planet over several weeks and months arises largely from its change in distance from the earth. Mars, for example, appears brightest when it is at *opposition*, or directly opposite the sun. It is then fairly close to us. It gets fainter as it moves away from the earth and disappears when it gets into the sun's neighbourhood. Finally, the fact that the sun, moon and planets all move in one particular part of the sky (a band or belt known as the *zodiac*) means that the solar system is a fairly flat affair. Or, as we say in astronomy, the planes of the orbits of the moon and planets are inclined at small angles to the plane of the earth's orbit.

So far all was well, but Copernicus made one big mistake. He clung to the old idea that the moon and planets travelled with unchanging speeds

Movement of a planet on the background of stars.

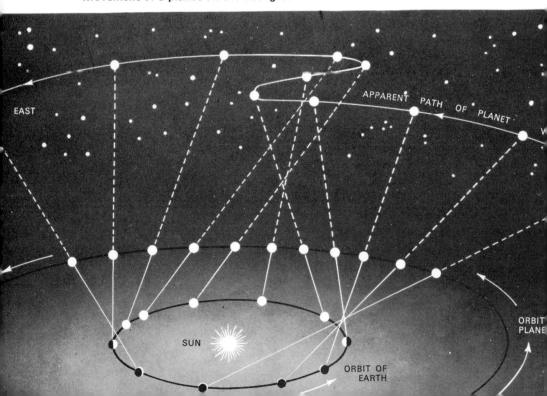

in circular paths. How was he to know that they moved otherwise? He also found that each body required two circles if theory was to agree with observation. When he got down to details he found that the whole thing got quite involved. He even had to move the sun slightly from the centre and put a point in its place. Small wonder, then, that few astronomers could follow all his arguments. Some of those who did, thought it all very clever but no more than an interesting exercise in geometry. Comparatively few, like Copernicus, believed that it represented the real thing—that the planets actually did travel round the sun, the earth among them.

Today, when the sun-centred view is taken so much for granted, we tend to forget how strange it must have appeared to people in the sixteenth century. Practically everyone then thought that the earth was at the centre of the universe, the more so since this idea had been generally accepted right from earliest times. Had not that great thinker Aristotle given good reasons why the earth should be at the centre? Did not all heavy things move towards the earth's centre? What direct evidence was there of a moving earth? If it did move, why wasn't the moon left behind? Instead of making things easier to understand the new theory made them more difficult. So it was not surprising that many leaders of thought strongly objected to it.

Copernicus died in 1543, the same year in which his great book on the new system was printed. Later in the same century Tycho Brahe, one of the observers of the supernova of 1572, was making regular observations of the positions of the sun, moon and planets. Using large quadrants and other instruments he made careful and accurate measurements over many years. After his death in 1601, his valuable records passed into the hands of Johannes Kepler, his assistant. Kepler began a long and difficult study of the material. It would, he hoped, show beyond any shadow of doubt that the Copernican theory was correct. But progress became more difficult the farther he went. The observations of Mars, for example, just would not line up with the idea that Mars moved with constant speed in a combination of circular orbits. He tried systems of two and more circles and even some oval ones but without success. Finally, in 1609, he hit upon the solution. Mars and the other planets didn't move in circles at all but in single ellipses. True, the ellipses were almost circles but they *were* ellipses and the motions in them were by no means constant.

Kepler expressed his results in the form of two laws:

(1) The orbit of each planet is an ellipse, with the sun at one of the two foci.

(2) The motion of each planet in its orbit is such that the line joining the sun to the planet (the *radius vector*) sweeps out equal areas in equal times.

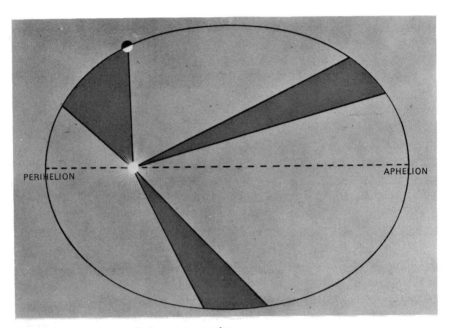

Kepler's second law of planetary motion.

The laws explained all the then-known irregularities in the movements of the sun, moon and planets. The first one did away with Ptolemy's idea of motion in two circles. The change in the apparent size of the sun, for example, is due to the fact that the earth's orbit is an ellipse. The earth is at *perihelion*, or nearest to the sun, early in January. Six months later it is at *aphelion*, or farthest from the sun. The difference between the two distances, 3 million miles, is quite small compared with 93 million miles, the average distance, so the ellipse is almost a circle.

The second law made the sun's apparent motion a direct result of the earth's actual motion in an ellipse. As its radius vector sweeps out equal areas in equal times, the earth moves faster than average at perihelion and slower than average at aphelion. This is why, as the Ancient Greeks discovered, the seasons have unequal lengths. Summer, for instance, has about 93 days and winter about 89.

Ten years later, in 1619, Kepler announced a third and final law:
(3) The square of the times which the planets take to make one revolution round the sun are proportional to the cubes of their mean distances from the sun.

This means that if a planet's average distance from the sun is four times as great as the earth's it will take eight years to make one revolution. If it is nine times as far away the time will be twenty-seven years ... and so on. So if astronomers can find any one distance in the solar system, they can determine all the rest. The period of revolution of a planet is easily obtained by observation. One just measures the time the planet takes to travel once round the sky from a star back to the same star again. The difficulty lies in getting a reliable value for any one distance.

A most important distance is that between the earth and the sun. We know that if we measure the sun's direction from the ends of a base line formed by the earth's equatorial radius, the difference in direction is only 8·79 seconds of arc (written 8"·79 for short). This value, known as the sun's *horizontal parallax*, is roughly equal to the angle subtended by a halfpenny nearly half a mile away. Early instruments and methods were quite unable to reveal anything as small as this. Astronomers like Hipparchus and Ptolemy thought the sun's parallax was 3 minutes of arc (written 3'·0 for short), a value adopted by almost every other astronomer, Copernicus included, right up to the time of Kepler. This made the sun's distance roughly 5 million miles, or over 18 times too small. Tycho's observations, however, showed Kepler that Mars had no appreciable parallax. This being so, and since Mars could come much nearer to us than the sun, the value of 3' of arc for the sun was obviously too large. Kepler therefore reduced it to 1' of arc and so made the sun's distance (and the size of the solar system) three times larger than it was before.

Kepler's three laws made possible much better predictions of the positions of the planets and the times of eclipses of the sun and moon. Later, they led Newton to his discovery of the 'inverse square' law of gravitation. All three, Newton showed, could be explained by supposing that each planet is controlled by a pull or force of attraction from the sun.

Also, that the force decreases as the square of the planet's distance from the sun increases. Armed with this 'inverse square' law he not only accounted for all the then-known movements of the moon and planets but even predicted some hitherto unknown ones. Further, he could find the relative masses (masses, that is, relative to the earth) of planets which, like Jupiter and Saturn, were accompanied by satellites. His discovery of the law was without doubt one of the greatest, if not the greatest, ever made in physical science. It underlies our knowledge of the mechanics of the solar system and system of stars. It describes the paths and motions of all bodies under the control of a gravitational force, whether they be rockets and projectiles, planets and comets, natural and artificial satellites, space probes, or stars moving in pairs, groups and clusters in space.

If the Copernican or sun-centred system is true it should be possible to detect a shift in the positions of the stars due to parallax. Suppose (as Copernicus did) that the stars are all at the same distance from the sun. Then as the earth travels round the sun it will be nearest to some of them at one time of the year and farthest from them six months later. We now know that the stars are at different distances from the sun. So as the earth swings from one side of the sun to the other a nearby star will appear in one direction at one time of the year and in a slightly different direction six months later. It should therefore move slightly in relation to the background of much more distant stars. Until recent times nothing of the sort was seen, for so distant are the nearest stars that their angles of parallax are extremely small.

Some early astronomers, Tycho Brahe among them, thought that the absence of any shift due to parallax was a strong argument against the Copernican system. Some suggested that the stars were too far away to show a parallax. Others rejected this. To them the idea of a great empty space between Saturn and the stars was quite absurd. Copernicus himself had little to say about the stars except that they were at enormously great distances from the sun. 'For what was proved', he wrote, 'is only the vast size of the Heavens compared with the Earth, but how far this immensity extends is quite unknown.'

Proof that the earth travels round the sun came in 1675, although in a rather roundabout way. Earlier that century Galileo had turned the telescope to the heavens and discovered, among other things, four of the moons or satellites of Jupiter. By 1675, thanks to improvements in the

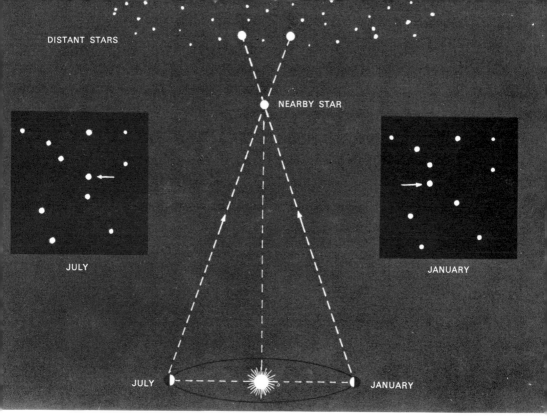

Stellar parallax.

telescope, astronomers had kept a close watch on these satellites. They could, among other things, see them regularly disappear into the shadow of Jupiter and then reappear. The times of these eclipses were recorded and it became apparent that they got more and more delayed when the planet moved away from the earth. The greatest delay occurred when Jupiter was on the other side of the sun or farthest from the earth. It became zero when Jupiter was on the same side of the sun or closest to the earth. Olaf Römer, a Danish astronomer, realised that this had something to do with light. He suggested that, contrary to the general opinion, light took time to travel from one place to another. The apparent delay in the times of the eclipses was therefore due to the time taken by light to travel over the additional space. 'Additional', that is, to the space between Jupiter and the earth when Jupiter was at its closest to us. His own observations showed that light took 22 minutes to cross a distance equal to the diameter of the earth's orbit. This meant that it had a velocity of

130,900 miles a second. The modern value, based on laboratory experiments, is 186,262 miles a second.

The first direct proof that the earth is moving round the sun was given by James Bradley. In 1725 Bradley set up a special telescope with the idea of trying to detect stellar parallax. With this venture he had no success whatever, but he did discover something equally important. This was, that once a year, stars well clear of the plane of the earth's orbit described tiny ellipses. In some respects the effect resembled what one would expect to get from parallax, but with one important difference—it was shared by all the stars. By 1728 Bradley knew that parallax played no part in the results of his observations. What, then, *was* the cause? The answer came quite unexpectedly while he was taking a trip on the Thames. He noticed that although the direction of the wind remained the same, the wind vane on the masthead changed direction with every turn of the boat. He realised that the change was an effect of relative velocity. The same kind of effect can be seen when we travel through heavy rain in a car or train. Although the raindrops are falling more or less straight down, they hit the side windows along sloping paths. The faster we go the more nearly horizontal do the splashes get. Bradley realised that in his case the speed of the wind corresponded to the velocity of light, while that of the boat corresponded to the velocity of the earth in its orbit. For any one star the greatest shift ($20''\cdot47$) from the true direction occurs when the earth travels in a direction at right angles to that of the star. It drops to zero when the earth travels directly towards or away from the star. Bradley called the effect the *aberration of light*. The value $20''\cdot47$, the *constant of aberration*, is equivalent to the average velocity of the earth divided by the velocity of light. Knowing the latter, the average velocity of the earth is found to be $18\frac{1}{2}$ miles a second.

After Bradley had discovered aberration he found that the stars still shared in another slight movement. Extremely careful observations over many years showed that this smaller or residual shift was tied up with the motion of the moon and also with the changing direction of the earth's axis.

We saw earlier that if the earth's axis is carried through the north pole, it meets the sky at a point called the north celestial pole. Also, that the bright star *Polaris* now happens to be fairly near this point. It hasn't always been like this. About 125 B.C. the Greek astronomer Hipparchus discovered that the stars were all slowly changing their positions relative to the *celestial equator* or projection of the earth's equator on the sphere of stars. The effect, known as the *precession of the equinoxes*, was studied by later astronomers but no one could track it down to its cause. It became clear, however, that the north celestial pole was a *moving* point—that thousands of years ago it was a long way from *Polaris* and would again be a long way from it in thousands of years' time. Newton, armed with his inverse-square law of gravitation, finally found the cause. The pulls of the sun and moon, he said, have a disturbing effect on the rotating earth. They act on the earth's equatorial bulge and give the earth a wobble like that of a spinning top. The top, we say, 'precesses'. So does the earth. But in the earth's case one complete wobble or precessional cycle takes, 25,800 years.

The earth, we saw, is like a gyroscope. This is true in the sense that the disturbing pulls of the sun and moon cannot cause it to tip over. It is not quite true if we mean that the earth's axis stays fixed in a constant direction. In a human lifetime, however, the change is so small that to all intents and purposes the axis is fixed. For us and for several generations to come, *Polaris* will remain the North Pole Star. But in 12,000 years' time, when *Vega* is the nearest bright star to the north celestial pole, the stars in the northern sky, *Polaris* among them, will all appear to travel round *Vega*.

What Bradley discovered was *nutation*, a slight regular nodding or to-and-fro-motion of the north celestial pole. This too is caused by the moon, or rather, by the moon's varying attraction on the earth's equatorial bulge. The discovery, one might think, was of no practical importance so why bother with it? Yet astronomy has been built up largely by discoveries of this kind. Unlike those of wireless waves and the electron they have led to no material benefits for mankind. Instead, their importance is like that of the pieces of a jigsaw puzzle. They help to make the picture complete. Bradley discovered such a piece—two in fact—and the story of his work is a fine example of scientific research. More often than not scientists set out to find one thing and, like Bradley, unexpectedly come across something quite different. How dull science would be if the expected always did turn up!

CHAPTER 4

The Telescope

MODERN astronomy owes most of its success to the telescope. Without it astronomers would know next to nothing about the surface of the moon, the nature of the planets, the different types of star, and the structure and size of the universe of stars. Strange to say we do not know for sure who invented the telescope. Some historians give all the credit to Porta, an Italian interested in all kinds of scientific inventions. Others favour the claims of certain opticians in Holland. The majority think it was Hans Lippershey, a spectacle-maker and glass-grinder of Middelburg who, towards the end of 1608, showed a telescope to members of the States-General, the governing body of the Netherlands. At any rate it now seems pretty definite that the telescope was invented in Holland sometime during 1608.

A story says that Lippershey came across the instrument purely by chance. An apprentice in his workshop happened to hold up two lenses some distance apart. He then looked through them and saw distant objects magnified. Lippershey tried to keep the matter a secret but the news soon reached other opticians. He also tried to get the States-General to appoint him sole manufacturer of telescopes. In other words, and to use a modern expression, he tried to 'corner the market'. When others tried to do the same, the States-General very wisely washed its hands of the whole affair.

By the end of 1609 telescopes were on sale in France, Germany, Italy and England. Galileo heard of them in May of that year. A correspondent in Paris wrote to say that a Dutchman had made an optic tube which magnified distant objects. Galileo was at that time professor of Mathematics at Padua and knew enough optics and geometry to realise how it was done. He mounted two lenses, one convex and the other concave, at the ends of a piece of lead tubing. When he looked through the tube with the concave lens close to his eye he saw distant things look as if they were near. His telescope magnified only three times but it was so new and exciting as to impress everyone who looked through it. Those who did, thought of its military uses or of the money it might make. Galileo thought of its value to astronomy. By early 1610 he had made telescopes which magnified up to thirty times and had used them to examine the moon, Jupiter and the Milky Way.

Galileo discovered that the moon's surface was not perfectly smooth as most of the Ancient Greeks had thought. Near the *terminator*, or line between the light and dark sides, he could see shadows cast by mountains and the outlines of curious ringed mountains called craters. The moon, it seemed, was another body like the earth, with mountains, valleys and, he thought, seas. Jupiter had a round shape and three very faint 'stars' near it. On the next night the three 'stars' had changed their positions in relation to Jupiter and also to one another. Six days later a fourth 'star' joined in. Observations over several weeks confirmed what Galileo had already guessed. Jupiter had at least four moons. How pleased Copernicus would have been had he known this! Here was a planet, moving in a far larger orbit than that of the earth, yet accompanied by four satellites. Here was another centre of motion far removed from the sun and the earth. Yet Aristotle had declared that everything travelled round the earth!

The telescope showed far more stars than could be seen with the unaided eye. The Milky Way, Galileo found, was made up of innumerable stars—so many that it was quite impossible to count them all. Saturn at first looked as if it were made up of three bodies. But as he watched it over many months the two side ones decreased in size and brightness and then disappeared. They then slowly returned, growing larger and brighter until Saturn looked oval, like an olive. Here was evidence of further change in the heavens, but just what was going on he could not tell. His telescopes were not powerful enough to show the true state of affairs—that Saturn was surrounded by a system of rings. This knowledge came in 1659 when the Dutch astronomer Huygens observed Saturn with telescopes better than those used by Galileo. Huygens thought that there was just a single ring but realised that it must be extremely thin compared with its width. Being so thin it disappeared from view when its plane was directed towards the earth. Eight years later, when the plane passed well clear of the earth, the ring had the 'open' appearance which had baffled Galileo.

The telescope showed that the planet Venus went through a cycle of phases in much the same way as did the moon. But with one important

Changes in the apparent shapes and sizes of Mercury and Venus.

difference—with the change in phase went a marked change in apparent size. When Venus was near *superior conjunction*, or on the side of the sun opposite to that of the earth, it looked small and round. Some ten months later, when it was near *inferior conjunction*, or almost between the earth and the sun and therefore fairly close to the earth, it looked large and had the shape of a thin crescent. As Galileo watched the changes he realised what was happening. The planet shone by reflecting sunlight. It moved round the sun inside the earth's path and therefore showed different parts of its sunlit side at different times. He assumed, correctly, that Mercury did the same and concluded that both planets 'necessarily revolve round the sun'.

In June, 1611, Galileo observed dark spots on the sun. They changed in size and shape and moved in such a way as to tell him that they were on the sun's surface. Here was something that Aristotle and his followers had not bargained for. The sun was not perfect after all but had a spotted face. Further, the way the spots moved, showed that it rotated on its axis. If a body as grand as the sun could do this, Copernicus's claim for a spinning earth was not so absurd after all.

Galileo, as we have seen, did not discover the telescope. Nor, as we now know, was he the first to see sunspots or the satellites of Jupiter. But these are relatively minor issues. His importance lies in another direction—in the way he used his own discoveries and those of others. To him they fitted into the picture roughly outlined by Copernicus, a picture of the universe quite different from the one painted by Aristotle.

Anyone interested in astronomy who is fortunate enough to possess a small telescope will soon want to get something larger. Astronomers early in the seventeenth century felt the same. The trouble was that to make telescopes magnify more they had to be made longer and longer. Galileo's largest telescope magnified about 30 times and was just a few feet long. Some of those used towards the end of the century reached lengths of 12, 30, and finally 150 feet. The wonder is that astronomers could point the long flimsy tubes to objects like the planets and keep them steady. Yet telescopes like these led to a number of important discoveries. Huygens discovered the ring of Saturn, and Titan, one of Saturn's moons. Cassini came across three more moons (Rhea, Tethys and Dione) and discovered that the supposedly single ring was composed of two rings separated by a gap now called the 'Cassini division'. Hevelius of Danzig studied the moon and in 1647 produced the first map of its main surface features.

The great lengths of early telescopes were necessary because of optical defects in the lenses used. A telescope which has a single convex lens for its front lens or object-glass cannot form a clear, sharp, image of the sun or moon. The image is slightly blurred and surrounded by coloured fringes. Both defects, however, are made less obvious if the lens opening or *aperture* is made smaller. Object-glasses were therefore made with long focal lengths but comparatively small apertures. They formed fairly sharp but faint images of distant objects. These images were then magnified by means of an *eyepiece* which usually consisted of a single convex lens or two slightly separated convex lenses fixed in a short tube.

This kind of telescope, known as a *refractor*, was popular in Newton's time. In an attempt to improve it, Newton made a number of experiments with light. One of these was his famous 'prism experiment'. In this he darkened his room and let in sunlight through a gap in the shutters. The light then fell on a glass prism which spread it out into a rainbow-coloured band known as a *spectrum*. Others before him had made a similar experiment, but he was the first to realise its great importance in understanding the nature of light. It told him that sunlight, or white light, consists of several differently coloured lights. Also that when white light is bent or refracted it is dispersed into various different colours.

The refraction and dispersion of sunlight by raindrops account for the form and colours of the rainbow. They also account for the coloured fringe seen round bright objects in ordinary refracting telescopes. Because of this Newton thought it was impossible to improve the refracting telescope. He therefore turned his attention to an alternative form in which the main image was formed by a concave mirror. This type, known as the *reflector*, appeared in three different forms named after their respective inventors, Gregory, Newton and Cassegrain. All three used a concave mirror to collect and focus light.

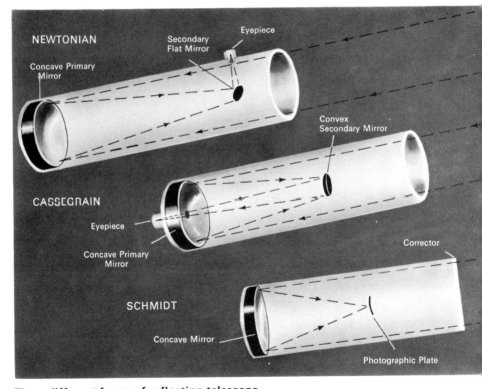

Three different forms of reflecting telescope.

A concave mirror has the advantage of producing images free from coloured fringes. Its reflecting surface, when given the shape or form of a paraboloid, also gives extremely sharp images. So once opticians and astronomers had mastered the art of grinding and polishing mirrors of this type, they could make telescopes quite large in aperture. In 1789, for instance, William Herschel completed a giant 48-inch reflector of 40 feet focal length. A large reflector is now found in nearly every major observatory. The largest so far is the Hale telescope of the Mount Wilson and Palomar Observatories, California. This has a mirror 200 inches in diameter, and when used along with other smaller mirrors, gives a choice of several different focal lengths.

The refractor also made good progress. In 1729 Moor Hall, a London barrister, invented a new type of object-glass which could give images as good as those given by a first-class mirror. Known as the *achromatic lens*, it turned the refractor into a compact and powerful instrument of research. Modern refractors are all achromatics. The largest is the 40-inch aperture telescope of the Yerkes Observatory, Wisconsin.

Large telescopes do not necessarily magnify more than small ones. Contrary to general belief, astronomers are not particularly interested in high magnifications. Much more important are large apertures, for these give bright images and also good *resolving power*, or the ability to reveal fine detail. Most of the objects studied by astronomers are extremely distant and therefore faint. They can be seen only in telescopes which collect a lot of light, that is, in large telescopes. They become all the more interesting if details of their form and structure can also be seen. Here again a large telescope scores over a small one. Suppose you looked at the moon through a small telescope magnifying about 500 times. It would look large but faint, with its features all rather blurred. Suppose you then looked at it through the 200-inch telescope on Mount Palomar, again with a magnification of 500. It would look no larger than before but so bright as almost to blind the eye. But if you used a dark glass as well you would see a breathtaking sight—craters, mountains, valleys, ridges and plains all so clear and sharp as to give the feeling that the moon was almost within landing distance.

In 1640, William Gascoigne, an amateur astronomer of Middleton, near Leeds, showed how the telescope could be improved still further. A spider very kindly spun a thread between two lenses, and in such a position that when Gascoigne looked through them at the sky, the thread looked like a thin black line. He saw at once that observations could be made with great accuracy if threads of this kind were mounted in the eyepieces of telescopes. For one thing, a telescope provided with two crossed threads in its eyepiece can be pointed accurately to a small object like a star. When the threads are used in this way we have what is called a *telescopic sight*. In another arrangement two threads are mounted in an eyepiece so that they are parallel to one another but capable of being separated by a known distance. In use they are moved until the object being studied just 'fits' between them. In the case of a planet they then form two parallel tangents. Their distance apart, read off from a finely graduated drum, is then converted into units of angle. An eyepiece of this kind is known as a *micrometer eyepiece* and is used for measuring small angular distances such as the diameters of planets and the separations of close double stars.

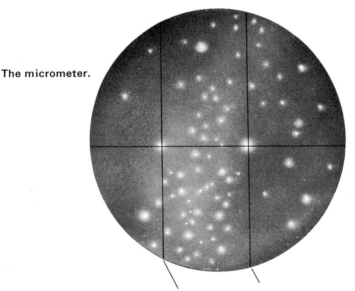

The micrometer.

MOVEABLE PARALLEL WIRES

CHAPTER 5

The Spectrograph

IF astronomers had to depend on the telescope alone they would still know very little about the stars. What, for example, are the stars made of? Are they all the same in brightness, size, temperature and average density? How do they shine and what is it like deep down inside them? To obtain information of this kind, astronomers rely heavily on photographs taken with the *spectrograph*, an instrument which is usually attached to a telescope at its eyepiece end.

The function of the spectrograph is to analyse light. It is based on Newton's prism experiment in which sunlight is split up into its various component parts or colours. In 1802, Wollaston, an English chemist and physicist, let sunlight into a darkened room through a narrow opening or slit. He then looked at the slit through a flint-glass prism and was surprised to see the spectrum crossed by seven dark lines. Unfortunately, he thought that the lines were only divisions between the colours. Twelve years later the German optician Fraunhofer repeated the experiment, this time using a narrower slit and a telescope with which to observe the spectrum. Much to his astonishment he saw several hundred lines. A few stood out much more clearly than others but all were dark, thin, and at right angles to the length of the spectrum. He then studied the light of other sources of light, but the results were puzzling. Venus showed all the strongest lines of the solar spectrum, the bright star Sirius showed a completely different set, and a sodium flame gave only two bright yellow lines on an otherwise dark background.

What did the lines mean? Why should they be dark in some spectra and bright in others? The solution came in 1859 when Bunsen and Kirchhoff, two physicists at Heidelberg, showed that the lines revealed the identities of chemical elements in the light sources themselves. Light from a glowing gas, they found, always produced bright lines. Different gases gave different lines, but any one gas always produced its own typical set. While glowing sodium gas gave two yellow lines, hydrogen gave four with the colours red, green, blue and violet. Light from hot solids or very dense gases, on the other hand, produced continuous bright bands. The colours in them ranged without a break from red through to violet, hence their name *continuous spectra*. Finally, when light from a hot solid was passed through a cooler gaseous element, the lines of the element appeared in the same places as before, but dark instead of bright. The element therefore stole or absorbed from the light passing through it just those parts which it emitted when it was made to glow.

The dark lines in the *solar spectrum*, or spectrum of the sun, therefore tell us that gases are absorbing some of the light from the sun's interior. The gases concerned, cooler than the highly compressed regions beneath them, form an envelope or atmosphere round the sun's main body. The remarkable thing is that these gases, all mixed up in the sun's atmosphere, make their own individual fingerprints in the form of spectrum lines. So to identify them the astronomer has only to match various sets of lines in the solar spectrum with corresponding sets produced by gaseous elements in the laboratory. The moon and planets shine by reflecting sunlight and therefore give solar-type spectra. That the stars have dark lines in their spectra shows that they too have gaseous atmospheres. No two stellar spectra are precisely alike but we know from the lines that the stars are made of similar substances. In fact, and although the physical conditions are vastly different, elements found on earth are similar to those found in the sun and stars.

As soon as the results of Bunsen and Kirchhoff's work became known, astronomers began to apply spectrum analysis to the stars. A great pioneer in this new and exciting field was William Huggins, an amateur astronomer who built his own private observatory in London. Between 1860 and 1869 Huggins examined the spectra of the sun, moon, planets and brighter stars. In 1864 he discovered that a misty patch in the constellation of Draco, the Dragon, gave a bright line spectrum. If it had been a very distant cluster of faint stars it would have shown a solar-type spectrum. The fact that it gave bright lines showed beyond all doubt that it was a mass of glowing gas. Photography was then beginning to play a part in astronomy and in 1872 Henry Draper, an American amateur astronomer, obtained the first photograph of the spectrum of a star. The star was *Vega* or *Alpha Lyrae* and the photograph or *spectrogram* showed four dark lines. This was a great advance, for it meant that astronomers

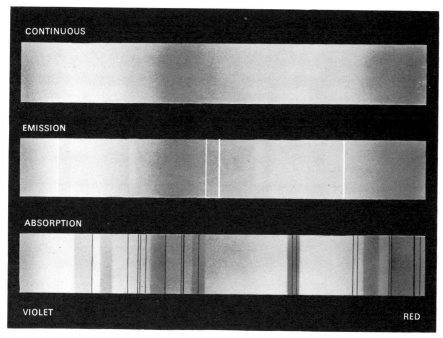

Three different types of spectrum.

could begin to study stellar spectra at their leisure instead of having to stay at the telescope all the time. Huggins also obtained a number of spectrograms, using for the purpose fairly compact spectrographs which he attached to the eyepiece ends of his telescopes.

Light enters a spectrograph through a narrow slit placed at the focus of an achromatic lens. It therefore meets the prism as a parallel beam. On leaving the prism the dispersed light is focused by a second achromatic lens to form a sharply-defined spectrum on a photographic plate. The part in front of the prism is called a *collimator*: that behind the prism is no more than a camera. Alternatively, the second or rear lens can also be used as the object-glass of a telescope. When this is done and an eyepiece is added the instrument is called a *spectroscope*. Usually more than one prism is used, for this increases the amount of dispersion and therefore the length of the spectrum. Dispersion can also be produced by a diffraction grating. This is a glass plate crossed on one face by a great number of evenly-spaced and parallel scratches of 'rulings'.

The spectrograph tells us not only what gases exist in the atmospheres of the sun and stars but also the temperatures and densities in those atmospheres. If, for example, the sun's atmosphere were made hotter than it is now, the familiar absorption lines would change in both intensity and number. In general, the spectra of extremely hot stars have comparatively few lines while those of cool stars have so many that they look like almost continuous dark bands. When the density is low the lines tend to be narrow, but they widen more and more as the density increases.

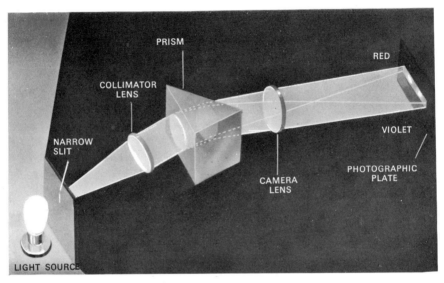

Principle of the spectrograph.

Thanks to the spectrograph we can detect some of the gases and vapours in the atmospheres of the planets. Since our observations are usually made from the earth's surface, many of the lines in the solar spectrum are due to absorption by oxygen and other gases in the earth's atmosphere. Fortunately these *telluric lines* as they are called, can be picked out fairly easily by observing them when the sun is high up in the sky (when they are relatively faint), and again when it is low down (when they are much stronger). In other words, the strength or intensity of the telluric lines depends on the sun's height in the sky and therefore on the thickness of atmosphere through which its light has to pass.

The spectrum of a planet like Mars can show even more lines, for sunlight, in being reflected by the planet's surface, travels twice through its thin atmosphere. The Martian spectrum therefore shows a combination of different sets of lines produced by absorption in the atmospheres of the sun, Mars and the earth. In the case of Venus, Jupiter and Saturn, sunlight penetrates only slightly into their dense atmospheres before it is reflected back to us. The difficulty is to pick out the additional faint lines. This is no easy task, especially if one particular element such as oxygen exists in the atmospheres of both the earth and the planet. Should this happen the two sets of absorption lines due to oxygen would coincide, those due to the planet being completely masked by those due to the earth.

One way of overcoming this difficulty is to send spectrographs to great heights by balloons, or better still, to mount them in artificial satellites. This raises further problems, for the instruments have to have high dispersions and therefore tend to be large and heavy. They also have to be operated by remote control and suitable arrangements must be made for their findings to be sent back to the controlling ground stations. Another way is to take high-dispersion spectrograms of a planet when it and the earth are either approaching or moving apart. When this happens the lines produced by the planet are shifted slightly towards the red end of the spectrum and therefore away from the telluric lines. The existence of water vapour on Mars was discovered in this way.

The shift of the lines towards the red is known as the *red-shift*. It occurs whenever a source of light travels away from us, or for that matter, whenever we travel away from a source. The important thing is that the source has a relative velocity of recession. When the motion is reversed to produce a relative velocity of approach, the lines all shift towards the violet end of the spectrum. In both cases the shift increases as the velocity increases. And in such a way that we can, by measuring the amount of shift, determine the velocity. The shift of the lines, known as the Doppler-Fizeau effect, is therefore of immense importance in astronomy. For one thing, it enables us to measure the velocities of approach or recession of distant luminous objects like stars, nebulae and galaxies. Velocities of this kind are called line-of-sight or *radial velocities*. For another, it can tell us how fast the sun and planets are rotating. As these bodies rotate, one edge or limb approaches us while the other recedes. So if the slit of a spectrograph is first pointed to one limb and then to the centre of the disc, the lines of one spectrum will be displaced in relation to those of the other. Measurement of the displacement leads directly to the line-of-sight velocity of that part of the limb concerned. Once the body's radius is known, it is an easy matter to calculate its period of rotation.

Light can be thought of as a form of energy which travels from one place to another in the form of waves. Motion of this kind is seen when a stone falls on water in a still pond. As soon as the stone hits the water a number of waves spread outwards or radiate in the form of circular ripples. In this way some of the energy of the falling stone is carried to the sides of the pool. The distance from the crest of one wave to that of the next is called the *wavelength*. In the case of light the waves are extremely small and radiate in all directions from the source. They also have electrical and magnetic properties, hence the expression (found in textbooks on physics) that light is the visible part of the *electromagnetic spectrum*. And a very small part too, for the range in wavelength over the entire spectrum is immense. It extends from very short gamma rays only a few thousand millionths of a millimetre in length to long radio waves many thousand metres long. In the visible spectrum a convenient unit of wavelength is the *Ångström* (written 1Å for short). This is equal to one ten thousand millionth of a metre or one ten millionth of a millimetre.

The visible spectrum extends from about 7000Å in the red down to 3,000Å in the violet. Directly below the violet is the ultra-violet, noted mainly for its effect on photographic plates, then come the penetrating x-rays. The latter overlap the even shorter gamma rays which can go down in length to about 0·03Å. Directly above the red end of the spectrum lies the infra-red, a band of radiation noted mainly for its heating effect. This in turn overlaps the short radio waves which extend upwards without a break to the very long radio waves of early radio broadcasting. Yet although the range in wavelength is so great, all electromagnetic radiation travels through space with the speed of light, namely, 186,262 miles a second.

Of the radiation from the sun that reaches the earth only a small part ever gets through to its surface. The rest goes mainly towards keeping the atmosphere warm. While this is important for life on earth it is a great handicap to astronomical observation. Gamma rays, x-rays and the ultra-violet up to about 3,000Å are either scattered or absorbed long before they reach the ground. The biggest thief in this region is a layer of ozone

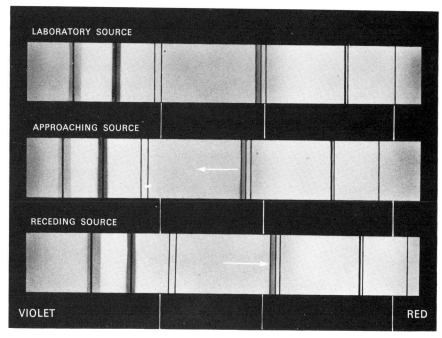

The Doppler-Fizeau effect.

in the upper atmosphere, but ordinary oxygen and nitrogen also take a heavy toll. Sunlight gets through without severe loss, except of course, when the sky is overcast. Another cut-off extends from about 20,000Å (near infra-red) to 1 cm. (*microwaves* or very short radio waves). The culprits in this case are mainly water vapour and carbon dioxide. Fortunately this region has a narrow but important opening between about 80,000Å and 120,000Å. Thanks to this, earth-bound instruments can be used to sample some of the infra-red coming from the planets, thereby providing a good indication of their temperatures. A third opening occurs in the radio section between 1 cm. and about 15 metres. Wavelengths longer than 15 metres are reflected back into space by the *ionosphere*, or layers of electrified particles in the upper atmosphere of variable depths, intensities and heights. There are therefore only three major openings or 'windows' in the atmosphere. Were it not for these we should, if we remained always on the ground, be completely cut off from the rest of the universe.

Most of the history of astronomy is an account of what has been discovered through the optical window. It says a great deal for the ability of man that he has learnt so much through so narrow an opening. Yet when he looks at a star through a telescope he receives only a tiny fraction of the energy the star sends us. Even when he analyses its light with a spectrograph and builds up a picture of the star itself he is still working with only a fragment of the available information. The rest is lost, either swallowed up or rejected by the earth's atmosphere.

The infra-red window clears considerably and becomes wider with increasing height above the ground. Mountain-top observatories are therefore better placed for making infra-red studies than those at or near sea-level. At a height of about 15 miles the window is so wide as to stretch from the near ultra-violet right through to the short radio waves. About 97 per cent. of the atmosphere, together with practically all the water vapour and carbon dioxide, lies beneath this level. One way of making

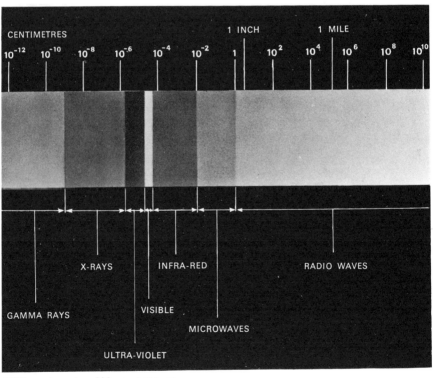

Different parts of the spectrum.

full-scale infra-red observations, therefore, is to use high-altitude balloons. This has now been done, and with particular success in an American project known as *Stratoscope II*. In this a giant balloon lifted a 36-inch reflecting telescope and other equipment to a height of about 15 miles. The telescope, fully controlled from the ground, was pointed to chosen objects while special infra-red detectors scanned their spectra. Successful scans have now been made of the spectra of Mars and Jupiter and of a number of red stars which emit strongly in the infra-red. The scans of Mars, made in March, 1963, revealed bands in the infra-red due to carbon dioxide and small amounts of water vapour in the planet's atmosphere.

Even at 15 miles the earth's atmosphere still cuts off the ultra-violet and lower wavelength regions of the solar spectrum. These are now being explored by means of high-altitude rockets. Although most of the work has been of an experimental nature it is now possible to record the main lines in the sun's ultra-violet and x-ray regions and also to pick up some of the sun's gamma ray emission. Ultra-violet spectrograms of many of the brighter stars have been obtained and a number of remarkably intense x-ray sources have been discovered.

Another way of studying the sun's short-wave emission is to place suitable instruments aboard artificial earth satellites. Two satellites have already done particularly well in this field—a British instrumented satellite (*Ariel I*, launched April 26, 1962) and an American orbiting solar observatory (*OSO I*, launched March 7, 1962).

The opening of the radio window was first made late in 1931 by Karl G. Jansky, a radio engineer of the Bell Telephone Laboratories. With quite simple equipment Jansky succeeded in picking up radio waves from parts of the Milky Way. His pioneer work was followed up in 1943 by Grote Reber, another American radio engineer, who built the first steerable dish- or bowl-type radio telescope. With this, Reber not only confirmed Jansky's discovery but also found that radio waves were coming from other parts of the sky. Those from the Milky Way were strongest in the direction of the constellation of Sagittarius but the others could not be associated with any particular objects. These discoveries marked the beginnings of radio astronomy, an exciting and powerful way of exploring deep space. How rapid and spectacular its progress has been will be seen in the following chapters.

CHAPTER 6

The Moon

Of all bodies in space, the moon was the first to give some idea of the immensity of cosmic distances. It did so in the second century B.C. when Hipparchus found that its distance was equal to $67\frac{1}{2}$ times the radius of the earth. His method, introduced in the previous century by Aristarchus of Samos and based on observations of eclipses of the moon, is no longer used. He did, however, introduce a method of his own which was decidedly modern in approach. Ptolemy used it some three centuries later and we have him to thank for its description. Owing to the size of the earth, the moon would appear in one direction when viewed from the earth's centre, and in a slightly different direction when viewed from a point on its surface. Unless, of course, that point happened to fall on the line joining the centre of the earth to the centre of the moon. In that case an observer there would see the moon at the zenith. An observer some distance away, however, would see the moon slightly shifted from the zenith. Ptolemy measured the shift, which was merely an angle of parallax. He also knew the distance between the two places and used this as a base line. Knowing the angle and the length of the base line, he calculated that the distance between the centres of the two bodies was 59 times the radius of the earth.

The modern trigonometrical method is similar to that used by Ptolemy. Astronomers at two widely separated observatories, one in the northern hemisphere and one in the southern, determine the position of the moon relative to the background of stars. Knowing the distance between the two observatories and the apparent change in the moon's direction they can determine its distance. Alternatively, an observer at one observatory can take advantage of the earth's rotation to give him his base line. Although this sounds simple enough in theory it is extremely difficult in practice. It is made easier if the two observatories are on the same meridian, and would be easier still if the moon were a point of light instead of a large area. The angle of parallax depends, of course, on the length

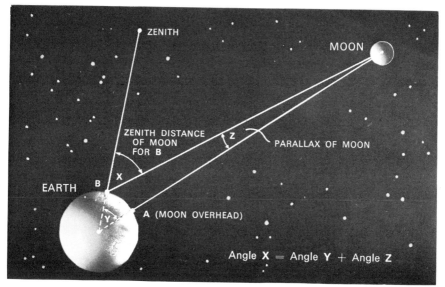

Ptolemy's method of finding the distance of the moon.

of base line used. Astronomers therefore refer to the moon's *horizontal parallax*. This is defined as the angular semi-diameter of the earth at its equator as seen from the moon. The moon's horizontal parallax (because of the elliptical shape of its orbit) varies between $53'\cdot 9$ of arc and $61'\cdot 5$. These extreme values correspond respectively to 252,710 miles (when the moon is at *apogee*, or most distant from us) and 221,463 miles (when it is at *perigee*, or closest to us).

Once the moon's distance is known, its size, 2,160 miles, can be easily calculated from its apparent diameter. This then leads to its density in the following way. Suppose the earth were connected to the moon by a straight but weightless rod. If the two bodies had the same mass they would balance like a dumb-bell at the middle of the rod. The mass-centre or centre of gravity of the earth-moon system would then be at the half-way point. Observations of the apparent motion of the sun reveal that it is not the centre of the earth but the centre of gravity of the earth-moon system that travels round the sun in an elliptical orbit. They also show that the centre of gravity is 2,903 miles from the centre of the earth. This is $\frac{1}{82}$ of the total distance between the centres of the two bodies. So if we divide this distance into 82 equal parts or steps, the centre of gravity will be one step away from the earth's centre and 81 steps away from the moon's centre. In other words, the mass of the moon is $\frac{1}{81}$ the mass of the earth. Studies of the earth's gravity show that its mass in tons is $6 \cdot 6 \times 10^{21}$ or 66 followed by 20 zeros. We know its size and therefore its volume, and by dividing mass by volume can derive its average density. This comes to 5·512 approximately, or 5·512 times the weight of an equal volume of water. By the same method the density of the moon is found to be 3·3.

The comparatively small mass of the moon gives further support to the belief that our satellite has practically no atmosphere. Any air molecules and atoms of gas would need a speed of only $1\frac{1}{2}$ miles a second to escape for ever from the moon. So if the moon once had a dense atmosphere, nearly all or perhaps all of it has now leaped away into space. On the earth, owing to the greater mass and therefore greater gravitational pull, the 'escape velocity' is 7 miles a second. Since comparatively few molecules reach this speed, the earth, unlike the moon, can hold on to its atmosphere. For the same reason a rocket could get away from the moon much more easily than from the earth.

As telescopes became larger and more powerful, knowledge about the moon increased. By the middle of the nineteenth century most astronomers were agreed that our satellite was completely different from the earth. True, it had no light of its own, and presumably, no heat of its own. The ashen glow of its unlit part came from the earth, that is, from sunlight reflected by the earth. Hence the modern term 'earthshine' for the appearance often known as 'the old moon in the new moon's arms'. The moon had mountains and valleys, but the dark areas or *maria*, once thought to be seas, showed every sign of being waterless plains. No clouds or mists ever hid the surface markings or softened their severe outlines. No twilight haze ever blunted the sharp 'horns' of the crescent moon. Indeed, our satellite appeared to have hardly any atmosphere. Whenever it hid or *occulted* a star, the star disappeared abruptly. The effect was most striking when planets like Jupiter, Saturn and Venus were occulted. The moon's edge or limb seemed to have no effect whatever on their brightness and shapes. A dense lunar atmosphere would have scattered and bent their light, so causing them to appear fainter and slightly distorted. Modern observations confirm these earlier findings, but this does not mean that the moon is completely airless. Even if it is surrounded by a gas some ten million million times less dense than the air we breathe this is still an atmosphere in the scientific sense.

The telescope shows that the moon always keeps the same side or face turned towards us. This means that it rotates once on its axis as it travels once round the earth. A lunar day is therefore equal to $27\frac{1}{3}$ days and daytime and night on the moon both last about a fortnight. Actually, as Galileo first noticed, we can see slightly more than half of the moon's surface over a period of time. This is because of two effects known as *librations*. One of these, *libration in longitude*, allows us to see a little of the far side at the eastern and western limbs alternately. This is brought about by the non-uniform motion of the moon in its elliptical orbit. True to Kepler's second law, it travels faster than average when it is at perigee and slower when at apogee. Its rotation time, however, stays constant. The rhythmic difference between the two brings about a corresponding to and fro swing of the moon's face. The other effect, *libration in latitude*, is due to the fact that the moon's axis, although always pointing in a constant direction, is not perpendicular to the plane of its orbit. As a result the moon's face appears slowly to nod, and in so doing allows us to see a little of the far side at the northern and southern limbs alternately. So although we see only one half of the moon's surface at any one time we can, thanks to the librations, extend it to about $\frac{4}{7}$.

Most of the detail on the moon is best seen near the *terminator*, or region between light and darkness. Here the sun is low down in the lunar sky, causing the mountains to cast long shadows and making even hills quite conspicuous. As the moon travels once round the earth the terminator sweeps twice across the moon's disc to show up mountain ranges, isolated peaks, craters, long canal-like valleys (termed *rilles*), and great walled plains, some with central peaks. At the terminator, high peaks, catching the light of a rising or setting sun, can appear like stars shining in the darkness. Galileo noticed this appearance and used it to estimate the heights of the lunar mountains.

At and near the time of full moon, strange bright streaks extend out like rays of white light from certain craters and walled plains. They cast no shadows, cross hills and valleys alike, and are associated with crater-like features which, like the rays themselves, brighten under a high sun. The finest ray system is that belonging to the large crater Tycho in the moon's south polar region. At and near full moon Tycho becomes a white oval patch and some of its rays extend a thousand miles. Seen through binoculars or a small telescope the moon then looks like an orange,

56 · BOOK OF ASTRONOMY

with Tycho marking the place where the stem joins the skin. The rays are presumably some kind of surface deposit shot out by the parent crater or by the disturbance that caused the crater. But why only a few craters should have rays while all the rest have none remains a complete mystery.

Early in the nineteenth century two German astronomers, Beer and Mädler, made a careful study of the moon with a small refracting telescope. Their results were published in 1838 in the form of a book and map. In their opinion, the moon was an airless wilderness with features fixed for all time. As a result most astronomers turned their attention elsewhere.

Systems of bright rays on the moon. The bright crater near the top or southern edge of the moon is Tycho.

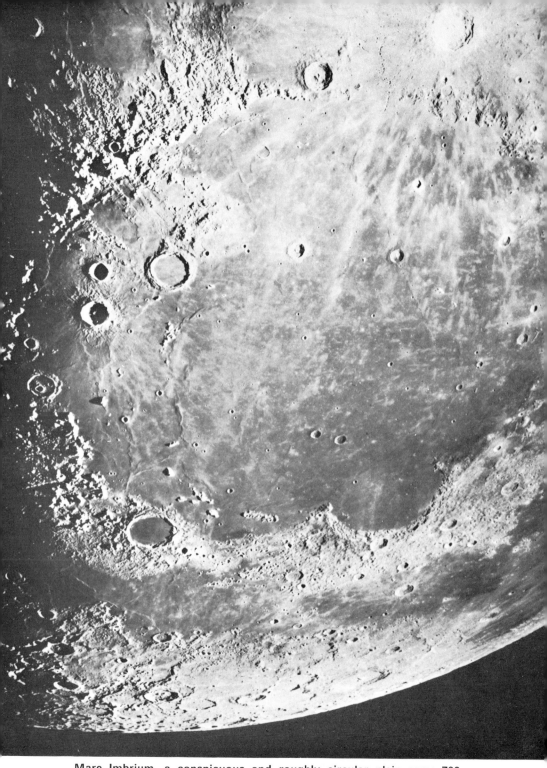

Mare Imbrium, a conspicuous and roughly circular plain some 700 miles in diameter. The crater at the top right is Copernicus.

There seemed to be no point in observing something that never changed.

Beer and Mädler were to some extent correct in their interpretation of the moon. No wind, rain, rivers or glaciers wear away its craters or fill in its valleys. They must look the same now as they did when they were formed millions and perhaps thousands of millions of years ago. But how were they formed? At one time it was widely held that the craters were all volcanic in origin. Yet although there are tens of thousands of them, comparatively few look like the craters of volcanoes on the earth. It is also difficult to see how the large ones, some 60 to 80 miles across, could have been formed by volcanic action alone. Another suggestion is that molten rock and gas, under great pressure, pushed up the semi-solid surface into huge swellings. These alternately collapsed and expanded, thereby building up the circular terraced walls we see today. Certain dome-like features scattered over the moon's surface seem to support this idea. They may be swellings which, owing to insufficient pressure from below, failed to 'burst' and form craters.

A third suggestion is that the craters were formed by *meteorites*, that is, by fast-moving lumps of metallic rock which came from outer space and crashed on the semi-plastic or perhaps solid surface. If so, meteorites must have been far more numerous in the past than they are now, for no new large craters have been formed in the last two centuries or so. Nor has any large-scale disturbance been seen which could have been caused by a meteorite. On the other hand we must remember that two hundred years is an extremely short time in the moon's past history. Even if one fairly large meteorite fell every thousand years the moon would have collected 4,000 sizeable craters over the past 4 million years. Yet according to modern estimates the moon's age is about 4,500 million years, or roughly the age of the earth. Another important point is that very large craters were not necessarily formed (if formed at all) by equally large meteorites. An object weighing several tons and travelling at some 30 to 40 miles a second would hit the moon with explosive force and form a crater much larger than its own diameter. Further, the crater would be circular (or almost so) even if the meteorite came in along an oblique path.

Some scientists think that even the maria were formed by meteorites, but since these are comparatively smooth regions, molten rock must have flowed in afterwards and filled them almost to the brim. The truth is that the more we find out about the moon's features the more difficult does it become to imagine that they were formed by only one process. Some craters overlap others. If the overlapping craters were formed by meteorites we should expect the original ones to be mere ruins, yet many show no signs of damage whatever. A number of craters on the earth are definitely known to have been formed by meteorites. But if the moon's craters were formed in this way why isn't the earth littered with craters? Perhaps they have been worn away over geological time, but one would still expect to find more traces of them than there are.

No two craters on the moon are exactly alike. Some are just a few hundred feet across, others could enclose a whole English county. Some have crater pits on their walls or enclose pits, craters, peaks and rilles.

Section through the lunar crater Copernicus.

While many craters on the maria stand out quite clearly, others, aptly called 'ghost craters', have rims so low and broken that they can be seen only when they are near the terminator. The craters certainly vary greatly in depth. A few have floors or interiors raised high above the general level of the surrounding regions. Others have deep floors and are like hollows in the surface. In fact, crater floors can fall anywhere between the two extremes, thereby making the puzzle of their origin all the more difficult. Perhaps all three crater-forming processes, volcanic, igneous and meteoritic were involved. But when, in which order and to what extent?

Just as geologists can build up something of a picture of the past history of the earth's crust, so scientists hope one day to do the same for the moon. The tantalising thing is that they cannot as yet walk over and drill into what they wish to explore. It's all over a quarter of a million miles away. At this distance the shape and structure of a feature about 1,500 feet across are difficult to see even in large telescopes used under the best possible observing conditions.

Astronomers can now study the moon with tiny but extremely sensitive heat detectors known as *vacuum thermocouples* or *radiometers*. When one of these is mounted at the focus of a large telescope it responds to what little heat the moon sends us. Work in this field shows that the temperature of the moon's surface tends to be very high in daytime and very low at night. This is to be expected. After all, having no atmosphere like the earth's, the moon has no protection from the intense heat of the sun in daytime and nothing to prevent the surface heat leaking away into space at night. The long daytime and equally long night add still further to the problems facing any would-be lunar explorer. Which would you prefer—nearly fourteen days in the glare and heat of the sun or the same time in the darkness and bitter cold of a lunar night? To live through both you would certainly need a well-insulated and pressurised space-suit.

Observations made with heat detectors show that under a high sun the temperature of the ground at and near the moon's equator can reach about 100°C., or the boiling point of water at the earth's surface. Yet in the same region far into the lunar night the temperature drops to about −200°C. Away from the equator the maximum temperature falls off rapidly to reach only −50°C. near the poles. At the poles the sun stays close to the horizon. Its low-slanting rays therefore have little heating power when they meet a fairly flat expanse of country. They also fail to reach the ground inside some of the deeper craters which therefore have their interiors plunged in perpetual darkness.

Another interesting discovery is the way the temperature changes during a total eclipse of the moon. As the moon enters the earth's shadow the temperature drops by about 200°C. to reach its lowest at the end of the eclipse. When the surface is clear of the shadow, the temperature slowly rises but does not reach normal until two to three hours later. It is also found that maximum temperatures occur about two to three days *after* full moon. There is therefore a delay in the heating and cooling of the surface. This can mean only one thing—that the surface is a poor conductor of heat. Laboratory experiments indicate that the observed changes in temperature are accounted for if we suppose the moon to be covered by a layer of fine dust. They do not, unfortunately, tell us how deep the layer is.

Dust could form on the moon in at least two ways—by the sudden cooling and heating of the rocks during lunar eclipses and by meteorites. The latter rain steadily on the earth and must therefore do the same on the moon. Here, all but the very largest and smallest (*micrometeorites*) are completely vapourised during their swift flight through the atmosphere. On the moon they all reach the surface at full speed, either in the form of dust or to make dust when they hit the rocks. The surface is also bombarded by electrified particles (*protons*) from the sun. According to the astronomer Thomas Gold, these give the dust particles a positive electrified charge. The latter then repel one another and hop about like Mexican jumping beans. If they are on a slope they will hop downhill under the influence of gravity. In this way, over the course of millions of years, the dust would tend to fill hollows and cover the maria, perhaps to a considerable depth. But if this is so, why do the maria look dark and have a patchy appearance? Through the telescope they look more like great lava plains than deserts of dust.

The moon can also be studied by radio telescopes adjusted or 'tuned' to receive microwaves. These are very short radio waves produced, in the moon's case, by the heat beneath its topmost layer. They therefore provide astronomers with information about regions below the visible surface. And for different levels too, for the longer the wavelength used the deeper the level that can be explored. From studies of this kind it appears that the temperature about three feet down is fairly constant at about $-30°C$. Whether it is lower or higher than this deeper down is not known for sure. A recent Russian report makes it higher. At any rate, the great changes in temperature on the moon are undoubtedly confined to its surface layers. They are, so to speak, only 'skin deep'. So to avoid these changes lunar explorers could establish their bases underground.

What is it like on the moon? Man will not know for sure until he steps upon its surface. He can, however, now form something of a general although very inadequate picture. On the moon, with no atmosphere to scatter sunlight, the sky always looks deep black. The stars can therefore be seen in daytime as well as during the night. The earth, four times the diameter of the moon, is seen to go through a monthly cycle of phases. When full, the light from its bluish-white disc is about fifty times brighter than moonlight. The sun is so brilliant that just to glance at it could blind the eyes. When low down in the sky (and just as bright as when it is high up) it makes the moonscape a scene of savage contrasts of light and darkness.

To anyone standing on the moon the skyline would be only two miles away. This is because the moon, about a quarter the size of the earth, has a more steeply curved surface. The low surface gravity (about one sixth that of the earth) means that a spaceman would have a feeling of lightness. If he weighed 12 stone on the earth he would weigh only 2 stone on the moon. It does not follow, however, that he would be able to jump 36 feet on the moon if he could just clear 6 feet in a high-jump on the earth. In the latter case the jumper raises his centre of gravity about 3 feet. With a similar muscular effort on the moon he should be able to raise it 18 feet.

In some parts of the moon the moonscape looks as if an army of giants had been at work, tearing open the surface, hurling mountains about and pouring great quantities of dust into the hollows. Elsewhere it looks severe but depressingly flat. Judged from the lengths of their shadows, some mountains are as high as the tallest mountains on the earth. Many have rounded shapes and from a distance look more like molehills than church steeples. At times they cast long tapering shadows as if to suggest that they are tall and slender, but this is only because the sun is then low in the sky. Under a low sun the craters look deep and impressive. In fact, most of them are shallow saucer-like depressions. The mountains forming the rim of a large one may rise some two to three miles above the general level of the ground, but for a crater having a diameter of 60 miles or more these heights are relatively small. Finally, the silence and stillness, the absence of life of any kind, and the harsh monotony of it all produce a scene of complete and utter desolation.

Does anything ever happen on the moon? A variety of minor changes have been reported but it is difficult to tell whether they were actual surface changes or effects produced by tricks of lighting or of the observer's vision. The most interesting one in recent years came in November, 1957, when the Russian astronomer N. Kozirev detected an unusual reddish haze over the central peak of the crater Alphonsus. At the time he happened to be taking spectrograms of this part of the moon. When he studied them a few hours later he was surprised to find that they contained lines additional to those normally present. These lines, it now appears, were due to glowing carbon vapour. Kozirev claimed that he had observed a volcanic eruption—that hot, molten rock had gushed out on the surface and sent up a cloud of dust, ashes and carbon vapour. Since then, other astronomers have looked in vain for signs of permanent change on the floor of Alphonsus. The general view is that a pocket of carbon vapour escaped to the surface where it was made to shine under the action of strong sunlight.

Late in 1961 Kozirev obtained spectrograms of the region of the crater Aristarchus which indicated the presence of a further disturbance, the vapour this time being molecular hydrogen. Two years later Greenacre and Barr, two astronomers at Lowell Observatory, Flagstaff, Arizona, saw three small red-coloured areas in the same region. The patches lasted many minutes but disappeared before photographs could be taken. The telescope used was a large refractor of 24 inches aperture, and Greenacre and Barr claimed that the patches would not have been visible in a much smaller instrument. Since then other observers have seen coloured patches inside Aristarchus, but what these patches were, remains a mystery.

Another way of finding out more about the moon is to send out small space vehicles called *lunar probes*. The first steps in this direction came in 1959 with the successful launching by the Russians of three *Luniks*. In January of that year *Lunik I* passed within 5,000 miles of the moon and then carried on to become an artificial planet. The information it sent back contained the first news that the moon had at the most an extremely weak magnetic field. In the following September *Lunik II* actually hit the moon, but in doing so its instruments went out of action. *Lunik III*, launched in October, travelled round the moon and obtained the first photographs of part of its far side. The cameras used were small and the pictures, televised back to the earth, lost much detail in the process. But a sufficiently large number of good ones enabled Soviet astronomers to build up a chart of about 70% of the far side. Most of the rest of the far side has since been photographed by *Zond-3*, a Soviet lunar probe which successfully completed its mission in July, 1965. Thanks to these achievements we now know that the far side is similar in nature, although not in general appearance, to the near side. It consists almost entirely of a continent of mountains and craters. The maria, few and far between, are all much smaller than those on the near side. The photographs also covered features on the near side—various maria and craters normally seen near the moon's limb and made to appear greatly narrowed or foreshortened.

Today the moon is being more closely studied than ever before. This is largely because the rocket has begun to put it well within man's reach. He has the means of getting there but not, as yet, of arriving in one whole piece. To ensure success in so great an undertaking he will also have to know what conditions he must face when he does get there. Hence the

64 · BOOK OF ASTRONOMY

importance to the American Apollo or 'Man-on-the-Moon' project of the brilliant success of *Ranger 7*. This lunar probe crash-landed on July 31, 1964, right inside a target area in the Mare Nubium. During the last thousand miles or so of the flight its cameras took 4,316 photographs, all of which were televised back to the control station in the United States.

The *Ranger 7* photographs showed surface details never seen before. In particular, those taken at heights below three miles showed clusters of small craters or pits on what was thought to be a smooth surface. The final picture, taken at a height of about 1,000 feet, and covering an area about the size of a tennis court, even showed pits only about 18 inches wide. Yet we still cannot say for sure what the moon's surface is like. The photographs throw no light on the old problem of how the larger craters were formed. One expert thinks that they indicate a hard surface with little dust on it. Another, that they show that the surface is soft and spongy.

The big problem with the *Ranger 7* photographs is their interpretation. This is also true of the additional thousands of photographs obtained by *Rangers 8* and *9* which crash-landed on the moon in 1965 on February 18 and March 24 respectively. The Americans therefore plan to remove all doubts by soft landing a *Surveyor* probe on the moon. Instruments carried by *Surveyor* could then radio back information about temperatures, ground hardness and structure, gas pressures (if any) and meteoritic impacts. Even then the information will concern only a small area of the moon. What about all the rest? Of course, one could always resort to taking 'pot luck' when landing there. But then the venture is too expensive and human life too precious to allow taking chances of this kind.

Part of the moon photographed from a height of 50 miles above its surface by Ranger 8.

CHAPTER 7

Five Naked-eye Planets

OF the five planets known since earliest times, Mercury is the most difficult to observe. This is to be expected since its average distance from the sun is only 36 million miles, or just over one-third the earth's distance. It therefore never wanders far from the sun in our skies. When it does become a naked-eye object it is always fairly close to the horizon and shines in competition with the bright twilight glow of the rising or setting sun. Yet despite these drawbacks it can, when conditions are favourable, almost equal *Sirius* in brightness and be surpassed only by Venus, Mars and Jupiter.

Mercury has a diameter of about 3,010 miles and is therefore half as big again as the moon. Small size, coupled with great distance, means that its phases are difficult to see in small telescopes. Galileo failed to see them but they showed up in 1639 in a more powerful telescope used by an Italian named Zupus. They were also seen, quite independently, by Hevelius of Danzig in 1644. Each cycle of phases takes 116 days, during which time the planet's distance from us changes by at least 72 million miles, or the diameter of its orbit. The phases are therefore accompanied by changes in apparent size. The disc is largest (13" of arc) at inferior conjunction, and smallest (5" of arc) at superior conjunction. On the latter occasion it usually passes above or below the sun, for the plane of its orbit is inclined at about 7° to that of the earth's orbit. At times, however, inferior conjunction occurs when it is at one of the nodes, or the two points where its orbit crosses the plane of the earth's orbit. It then passes directly between us and the sun to appear as a round but tiny black spot on the sun's bright face. These passages across the sun are known as *transits*.

From calculations based on Tycho's observations, Kepler found that Mercury would transit on November 7, 1631. He did not live to see the event but Gassendi of Paris, forewarned well in advance, kept watch and saw it take place on the very day that Kepler had predicted. Many transits have been observed since then for they can occur at the rate of 13 or 14 a century. Five more are expected this century—May 9, 1970; November 10, 1973; November 13, 1986; November 6, 1993; November 15, 1999.

When Mercury is observed with fairly large telescopes, various hazy dark markings can be seen. They have been likened to the maria seen on the moon with the unaided eye. Schiaparelli, Antoniadi and other skilled observers have compiled charts of Mercury which show a number of named dark patches, spots and streaks. These features, like those of the moon, appeared to be fixed, so observers naturally concluded that Mercury rotated once on its axis as it travelled once around the sun. One hemisphere was thought to have perpetual sunshine and the other perpetual night.

In 1964, however, the giant 1,000-foot radar-radio telescope at Arecibo, Puerto Rico, was used to bounce short-wave radio signals (p. 95) off Mercury. The reflected signals or echoes carried definite traces of the Doppler-Fizeau effect due to the planet's rotation. But on analysis, and much to everyone's surprise, they indicated that the length of Mercury's day was 59 ± 5 of our days, and that the rotation was direct, or from west to east like the earth's. This result, needless to say, has not been accepted by many European astronomers.

Mercury's orbit is decidedly eccentric or oval-shaped. The distance at perihelion is 29 million miles and at aphelion 43 million miles. In consequence the change in orbital velocity (in accordance with Kepler's Second Law) is correspondingly great, being 36·5 miles a second at perihelion and 24 at aphelion.

A planet only slightly larger than the moon cannot be expected to have a dense atmosphere. Mercury has no satellite and is so tightly held by the powerful attraction of the sun that its disturbing effect on the motions of the other planets is very slight. Its mass is therefore only roughly known, but the value usually given is 0·054 of the earth's mass. If this is so the velocity of escape from its surface is only 2·62 miles a second. Under the sun's intense heat molecules would move at speeds faster than this and so leave the planet forever. Yet the appearance of unusual hazy patches and certain local changes in the sizes and clearness of the markings could mean that clouds of dust are being stirred up in a thin atmosphere. On the other hand the changes are so difficult to see that some astronomers doubt their existence altogether. The general opinion is that the planet is almost if not completely airless.

This conclusion is supported by estimates of the heat radiated by Mercury. In 1923-4 Pettit and Nicholson studied the planet with a radiometer mounted at the focus of the 100-inch reflector of the Mount Wilson Observatory. They found that the maximum temperature on the sunlit side was about 400°C. Hardly any radiation was received from the night side, so the temperature there was near absolute zero (-273°C.). Subsequent studies of a similar nature largely support these findings, thereby making it pretty definite that Mercury is at once the hottest and coldest of all the planets.

Venus, second planet from the sun, moves in our skies in much the same way as Mercury does But since its average distance from the sun is 67 million miles, or about twice that of Mercury, it can swing away from the sun by about 46° compared with 23° for Mercury. It therefore has the advantage of appearing in a well-darkened sky. Like Mercury it goes through a cycle of phases but in the longer period of 584 days. The change in phase is associated with a big change in apparent size. At superior conjunction, when the planet has a full disc, the apparent diameter is 11" of arc. Near inferior conjunction, when only a slender crescent is seen, it increases to about 67" of arc. The phases can therefore be seen in a telescope of moderate power.

Transits of Venus are fairly rare, but when they do occur Venus can be seen by the naked eye as a round black spot—provided, of course, that a dark glass is used. The earliest observed transit, in 1639, was seen by two English amateur astronomers, Jeremiah Horrocks and William Crabtree. The next two, in 1761 and 1769, had a special interest, for in 1679 Halley showed how they could be used to find the sun's parallax. His method involved observing the sun through telescopes at two or more widely separated stations and noting the times when Venus first appeared on the sun and then when it was about to leave. But owing to various causes it was almost impossible to estimate the required times accurately. Astronomers therefore disagreed among themselves and the results (8"·5 to 8"·8 of arc) inspired little confidence. The method was tried out again at the transits of 1874 and 1882, but again with no great success. The next two will occur on June 8, 2004, and June 6, 2012.

Venus is the brightest object in the sky apart from the sun and moon.

So bright, in fact, that it can sometimes be seen in broad daylight. This is partly because it is fairly large, its diameter of 7,650 miles making it over twice the size of Mercury and almost equal to the earth. It is also our nearest neighbour among the planets, being less than 25 million miles away at its closest. Further, it is so good a reflector of sunlight as to reflect some 59 to 67 per cent of the light it receives. This is because it is surrounded by a cloud-laden atmosphere. The clouds form a brilliant white blanket so dense that it is impossible to see what lies beneath.

That Venus has atmosphere is plainly evident from observation. When about to transit the sun its black disc is surrounded by a ring of light produced by the refraction of sunlight in the planet's atmosphere. Also, when it has the form of a thin crescent the horns or cusps of the crescent extend slightly farther than they would do if the planet had no atmosphere. In addition, when Venus occults a bright star the star rapidly fades before its light is completely cut off. All told, observation suggests that the atmosphere is dense and has a depth of some 50 to 60 miles.

Studies of the spectrum of Venus indicate that the atmosphere above the clouds is fairly rich in carbon dioxide. Although the amount is sufficient to smother earthlike creatures it doesn't follow that it is equally abundant beneath the clouds. There is also a small amount of water vapour. Its presence was first announced in February, 1964, by the American astronomer John S. Strong and his colleagues, who sent an infra-red spectrograph by balloon to a height of over 16 miles. Venus was then approaching the earth, so the bands due to water vapour in its atmosphere were slightly shifted from the telluric lines (p. 46). Unfortunately, it is difficult to tell from this how much water vapour exists beneath the clouds, but the amount is thought to be extremely small.

In the telescope Venus reveals only occasional dusky markings on an otherwise white disc. They are hazy and short-lived and must therefore be atmospheric features. More definite markings in the form of streaks and bands appear on photographs taken with plates sensitized to ultra-violet light. Since this radiation has comparatively little penetrating power the features recorded must also be atmospheric and exist among or above the white clouds. They too are fairly short-lived and therefore of little use when it comes to trying to determine the planet's period of rotation.

When a planet shows no permanent surface markings, it is extremely difficult to determine its period of rotation and the direction of its axis of rotation. Estimates of the period based on changes in the positions of the dusky markings range from 22 hours through to 225 days. Needless to say they are all unreliable. Spectrographic studies have failed to reveal any signs of a Doppler-Fizeau effect due to rotation, so the period is probably a long one.

Much the same could perhaps be said of recent American radar observations, but they are at least consistent. As in the case of Mercury, radar signals were bounced off Venus and examined for signs of the Doppler-Fizeau effect. Two quite independent sets of observations, made at Arecibo in 1964 and at Goldstone, California, in 1965, gave a rotation period of the order of 250 days in an east to west or retrograde sense. The reflected signals also showed that some parts of the planet's surface were better reflectors of short radio waves than others.

If Venus keeps one side continually facing the sun we could expect a great difference in temperature between the day and night hemispheres. Microwave studies, however, indicate that the difference is relatively small. They also reveal a startling overall temperature of about 300°C., although the results depend to some extent on the wavelengths used. That the temperature on Venus is of this order became known in 1963 after the data obtained by the American probe *Mariner 2* had been analysed. After a flight of 109 days this little space laboratory passed by Venus at a distance of 21,648 miles. As it did so a radio telescope or radiometer automatically scanned both the bright and dark hemispheres. The results were then radioed back to earth over a distance of 36 million miles. A number of allowances had to be made before the data could be changed into temperatures. The planet's surface temperature came to about 426°C., with a temperature difference of no more than 45°C. between the light and dark hemispheres. The scans also showed that the origin of the high temperature was heat trapped beneath the clouds and not, as some had suggested, by electrical activity in an ionosphere above the clouds. Venus is therefore a most efficient greenhouse, the more so since carbon dioxide tends to trap heat. Sunlight passes through the atmosphere and clouds to heat the surface, but most of the heat radiated by the surface cannot escape into space. So well does it get 'bottled up' and distributed in the lower atmosphere that the temperature remains much the same whether it is daytime or night.

The radiometer on *Mariner 2* also detected a spot in the cloud layer which was some 20° cooler than its surroundings. One suggestion is that it was caused by a very high mountain or mountains. But we really have no idea what the surface is like, whether rough or smooth, partly covered with chemicals in liquid form or completely dry. Only that it is hot—too hot for any form of life known on earth. There is probably hardly any water vapour and certainly no liquid water. So Venus could be a barren desert, with most of its rocks worn fairly smooth by dust storms swept along by tremendous winds. Indeed, one theory suggests that the high surface temperature is produced by friction between the dust clouds and the solid surface.

Since Venus has no satellite, astronomers have had to determine its mass by making careful studies of its motion. *Mariner 2*, however, led the way to a new and better determination of the mass. Radio techniques enabled the probe to be tracked with great accuracy over distances of many millions of miles. As it came near Venus the planet pulled it off its course and increased its speed. Calculations based on these changes have given Venus a mass 0·8150 times that of the earth. The probe also gave a more accurate value of the moon's mass, for the moon also deflected it by a known amount.

If we imagine the sun to be a globe 2 feet in diameter, Mercury will be a grain of mustard seed 82 feet away, Venus a pea at a distance of 142 feet, and the earth another pea at a distance of 215 feet. A model of this kind, set out in a field, cannot fail to bring home the extreme smallness of the planets compared with the sizes of their orbits. Inside the earth-circle, 430 feet in diameter, there are only a grain of mustard seed and a pea. Yet our model is far from complete for there are six more model planets to add. By the time we come to Pluto, the most distant planet, we shall require a circle about $3\frac{1}{4}$ miles in diameter and therefore a very large field indeed.

On the scale of the model, Mars, the fourth planet, is a rather large pin's head at a distance of 327 feet from the sun globe. The real planet has a diameter of 4,220 miles and its mass is only about one-tenth that of the earth. We can therefore expect it to be cooler than the earth and to have a fairly thin atmosphere. That the atmosphere is relatively thin is clearly shown by the telescope. Various well-defined dark markings can be seen. They are permanent enough to be surface features and enable the period of rotation (24 hr. 37 min. 22·6 sec.) to be fixed with great accuracy. They also show that the plane of the planet's equator is inclined 25 degrees to the plane of its orbit. So in these two respects Mars is similar to the earth. All places on its surface get a share of the sunshine and those in mid-latitudes have the seasonal changes in the sun's height which we on earth associate with spring, summer, autumn and winter. But with one important difference—Mars takes about two years to go once round the sun, so each 'season' lasts about six months.

Since Mars has a fairly elliptical orbit its distance from the sun varies by about 26 million miles. It appears at its brightest and largest at opposition, but unfortunately oppositions occur at intervals of 780 days or so. The most favourable are those which occur when the planet is at or near its perihelion. Even then it is still 34·5 million miles away and its disc is only 25" of arc in diameter. The last favourable opposition was 1956 and the next will not be until 1971. If, as in 1948 and 1956, opposition occurs when the planet is at aphelion, its distance can be over 60 million miles.

Mars shows three distinct features: two bright white polar caps, the dark markings already referred to, and large red-yellow areas which give the planet its characteristic reddish colour. The polar caps change considerably in size with the seasons, being largest in winter and smallest in summer. As on earth, but to a much greater extent, the north cap is large when the south cap is relatively small, and vice versa. Spectroscopic observations show that the caps are deposits of frozen water, so they may consist of layers of snow only an inch or two thick. They must be thin since they shrink so much and so rapidly as the weather gets warmer. In fact, the south cap sometimes vanishes completely. If this ever happened to the earth's polar caps many countries would be almost completely flooded by the sea. Oddly enough, any snow on Mars would sublime rather than melt as the temperature rises. In other words, it would pass directly into water vapour without first turning into liquid water. This is because of the thin atmosphere and therefore low atmospheric pressure.

The dark markings also change with the seasons, but in a more complicated way than the polar caps. A few dark areas hardly change at all. The rest gradually darken as the polar cap in their particular hemisphere gets smaller. The overall effect is as if a wave of darkening were moving from the polar cap towards the equator. The best explanation so far is that the changes are produced by the growth of some form of vegetation.

As winter changes into spring, winds carry moisture from the shrinking polar cap towards the equator to stir the plants in their paths into activity. This interpretation is supported by spectrographic studies. In 1958 W. Sinton of the Lowell Observatory, Arizona, obtained spectrograms of Mars which showed bands in the infra-red characteristic of many organic compounds. While this does not prove that life exists on Mars it is certainly a strong point in favour of some form of plant life.

The reddish areas form the greater part of the planet's surface. Sometimes they look featureless, at other times (mostly during a Martian summer) they contain dusky streaks and patches which often appear as extensions to the darker markings. No mountain ranges can be seen but any hills or fairly high mountains would not be seen anyway owing to the planet's great distance from the earth. Sometimes whole areas are blotted out by a yellow haze which slowly clears to reveal the dark markings again. This happened during the favourable opposition of 1956 when the dark markings were clouded over for weeks on end. The haze is thought to be composed of swirling clouds of dust swept up from the reddish areas. If so the latter are great deserts of red and yellow dust. This is additional support for the idea that the dark markings are areas of vegetation, for only living things could push their way up through the dust layers. White clouds are sometimes seen. They appear to be high up and are probably made of ice crystals. They are nothing like as widespread and dense as the yellow ones produced by the dust storms.

Optical studies indicate that the Martian atmosphere is extremely thin—so much so that the pressure at ground level is believed to be between about 7 mm. and 20 mm. of mercury. On earth at mean sea level the standard barometric pressure is 760 mm. of mercury. Spectrographic studies show that the atmosphere on Mars contains carbon dioxide and also water vapour, but so little of the latter that if it were all turned into liquid water it would form only a thin film over the planet's surface. There is probably plenty of nitrogen, but unfortunately this gas cannot be detected by the spectrograph. Oxygen has not yet been found so if any is present it must be extremely small in amount. The red colour of the deserts suggests that any oxygen the planet once had is now locked up in the surface rocks in the form of iron oxide or rust. Temperature estimates using vacuum thermocouples indicate that even at the equator with the sun high in the sky the temperature is only about 30°C. At

Mariner 4 close-up photograph of part of the Martian surface. The area covered is about 170 miles by 150 miles.

midnight, owing to the thin atmosphere, it must fall well below zero. Conditions like these would be fatal to most plants on earth but some, such as lichens, might survive.

About sixty years ago the American astronomer Percival Lowell, using refractors of 18 and 24 inches aperture, claimed that Mars was crossed by a network of straight lines which, he thought, were strips of vegetation growing alongside canals. Since then observers of Mars have continued to record long dark streaks, although it was generally realised that the streaks could not possibly be artificial waterways. They form no definite geometrical pattern and are nothing like as well-defined and straight as Lowell made them out to be.

If canals existed they would have appeared on the close-up photographs taken in July, 1965, by *Mariner 4*. As this little space probe flew within 9,000 miles of Mars its camera photographed sections of a long narrow strip of the planet's surface. The main features on the photographs were not canals but craters, dozens of them, all remarkably similar in appearance to those on the moon. They range in diameter from 3 to 75 miles and point to an inactive and incredibly old lunar-type surface.

Mars has two satellites, discovered by the American astronomer Asaph Hall in 1877 and named Phobos and Deimos. Phobos, only 10 miles in diameter, makes one circuit of Mars in 7 hr. 39 min. at a distance of 5,800 miles from the planet's centre. It therefore revolves roughly three times faster than Mars rotates and in consequence travels across the Martian sky in a west-to-east direction. Deimos, 5 miles across, has a period of 30 hr. 18 min. and is 14,600 miles from the planet's centre. Since the period is nearly equal to the Martian day, Deimos moves very slowly across the Martian sky in an east-to-west direction.

Leaving Mars and continuing outwards from the sun we come to mighty Jupiter, a planet some 318 times as massive as the earth and so large that it could contain over 1,300 bodies each the size of the earth. In a telescope of moderate power Jupiter shows a bright golden disc crossed by bands or belts of varying shades arranged parallel to the planet's equator. The disc has a distinct flattened shape, the distance between the poles being less than that across the equator. Measurements confirm this, for whereas the polar diameter is 82,800 miles the equatorial is 88,760 miles, a ratio of about 14 to 15. That an equatorial bulge of this large extent is produced by rapid rotation is shown by observation. By means of markings on the belts the period of rotation for the equatorial regions can be fixed at 9 hr. 50 min. 30 sec. The speed of the markings is therefore about 27,000 miles an hour, far greater than the equatorial speed on earth. Away from the equator, however, the period of rotation is about 9 hr. 55 min. If the markings were fixed to a solid body like the earth or Mars the period of rotation would be the same for all latitudes. Those on Jupiter must therefore belong to something less substantial such as an atmospheric mantle.

The way the markings change in appearance and position shows beyond all doubt that they are clouds or cloud-like features and that the belts represent great 'trade winds' or atmospheric currents which flow parallel to the equator. One large marking, known as the Great Red Spot, is fairly permanent. In 1878, after being pale pink and indistinct, it developed into a prominent red-brick oval patch some 30,000 miles long and 7,000 miles wide. Since then it has varied quite a lot in both colour and prominence. In 1957-8, for example, it was quite conspicuous yet in 1959 it vanished completely. It has also drifted away from its average position, the drift in latitude being several thousand miles and that in longitude as much as 400,000 miles, or much more than Jupiter's entire circumference. The fact that it wanders in this way suggests that it floats in the atmosphere. On the other hand it must have great depth, otherwise it would not have lasted so long. It may even be due to some disturbance on the planet's surface but owing to the dense atmosphere we have no direct knowledge of what this is like.

Spectrograms of Jupiter show that its atmosphere contains gaseous ammonia (NH_3) and methane (CH_4). Since both gases are compounds of hydrogen it is reasonable to conclude that hydrogen is extremely abundant. This is supported by theory. Although the planet is massive it is so large that its average density is only 1·33 times the density of water. Calculations based on the amount of polar flattening show that the density is much higher than average near the planet's centre and lower than average in its outer layers. The latter must therefore consist mainly of light gases like hydrogen and helium.

According to modern theories Jupiter is a decidedly chilly planet with little heat of its own. Recent radiometric measurements made with the 200-inch telescope show that its surface temperature averages $-145°$ C., which is no more than we might expect in view of its great distance from the sun. The surface in question, of course, is not that of Jupiter itself but one formed by the cloud layers. Beneath the clouds the temperature may be much higher.

In 1924 Harold Jeffreys was led to picture Jupiter as consisting of a rocky core surrounded by a layer of ice several thousand miles thick, which in turn was covered by an extensive atmosphere. Rupert Wildt, a German astronomer, drew up a similar picture, but in 1931 William Ramsey stated that Jupiter consisted mainly of hydrogen. Ramsey's theoretical work suggested that most of Jupiter's mass is in the form of a great core of metallic hydrogen 76,000 miles in diameter. Surrounding this is a shell of solid hydrogen 5,000 miles deep, above which is a comparatively shallow atmosphere of hydrogen gas and its compounds. At the low temperature of the atmosphere, however, much of the ammonia would be frozen to form clouds of crystals. Ramsey's model of Jupiter, now generally accepted, reminds us that hydrogen is an element rather than a gas. Under very great pressures such as those which could be found on Jupiter it can exist as a solid and even as a metal.

In 1892 the American astronomer E. E. Barnard discovered that Jupiter had a fifth satellite. The object is so faint and moves so near the planet that it can be seen only in large telescopes. Since then seven others have been found, so making twelve altogether. The four discovered by Galileo show definite discs in large telescopes and can therefore be measured with a micrometer. Two of them, Ganymede and Callisto, are respectively 3,000 and 2,800 miles in diameter and therefore larger than the moon. The other two, Io and Europa, have diameters of 2,000 and 1,750 miles respectively. The rest show no discs, so their sizes have to be judged from their brightness alone. They are thought to range in size from about 10 to 100 miles. The four outer satellites differ from the rest by having *retrograde* motions. That is, they move in a direction opposite to that of the planet's rotation. For this reason, and also because of their small size, they are thought to be 'captured' bodies. Long ago they probably travelled round the sun as minor planets but, getting too close to Jupiter, were 'caught' by its great gravitational pull. But if small size alone is any guide in this matter the other four small satellites also qualify as captured bodies.

When the Ancient Greeks arranged the planets in order of distance from the earth they always gave Saturn the largest orbit. Copernicus did the same in his sun-centred system, for Saturn moved slowest of all the then-known planets. The planet takes $29\frac{1}{2}$ years to orbit the sun, or over twice the period of Jupiter, and its average distance is now reckoned to be 886 million miles. In size it comes next to Jupiter, and like Jupiter, it is markedly flattened between its poles. The polar diameter is 67,200 miles and the equatorial 74,160, a ratio of about 11 to 12. So if we represent the sun by a ball two feet across, Jupiter is a moderate-sized orange a quarter of a mile away, and Saturn a small orange at a distance of two-fifths of a mile.

The telescope shows that the ball of Saturn is crossed by a series of parallel belts, but unlike those of Jupiter they show little detail. Bright spots sometimes appear. They are usually short-lived but show that the period of rotation at the equator is 10 hr. 14 min. and increases with increasing latitude. Spectrograms of Saturn are similar to those of Jupiter except that the lines due to ammonia are weaker while those of methane are stronger. The parallel markings must therefore be cloud belts, but at temperatures so low that most of the ammonia is frozen out of the atmosphere. Radiometer observations confirm this, the estimated temperature of the cloud surface being in the region of $-155°$ C. Saturn, it seems, is another planet with an abundance of hydrogen in its atmosphere and little or no heat of its own. Most of its material must be compressed into a dense core which probably consists largely of metallic hydrogen. If so the core must be fairly small owing to the amount of the equatorial bulge and the high rate of rotation. And for another reason too—Saturn has an average density of 0·71, or less than that of water. It is therefore so light that it would float on water, providing, of course, one had a big enough bowl.

The glory of Saturn lies in its remarkable system of rings. There are at least three separate rings, each thin, flat, and lying in the plane of the planet's equator. Cassini discovered two of them in 1675 when he saw the division named after him (p. 39). Much later, in 1850, G. P. Bond in America and W. R. Dawes in England independently discovered a faint inwards extension of the inner ring. It is known as the Crêpe Ring because of its dusky appearance, and is so transparent that the body of the planet can be seen through it. The other two rings are also partially transparent, for on the rare occasions when they pass over a bright star the star remains visible, although much reduced in brightness. That they cannot be solid or liquid was first demonstrated in 1859 by the Scottish scientist J. Clerk Maxwell. Using mathematical arguments Maxwell showed that rings of this nature would soon break up under the action

Jupiter, showing the Great Red Spot. 200-inch telescope.

Saturn. 100-inch telescope.

of gravitational forces. He then went on to prove that they must consist of swarms of tiny satellites or moonlets, for only then could they remain fairly undisturbed over great periods of time. Direct proof by observation came in 1895 when James Keeler at the Allegheny Observatory obtained spectroscopic evidence of the rotation of the rings. Small shifts in the spectrum lines due to the Doppler-Fizeau effect told him that the inner parts of the rings were rotating faster than the outer. If they were solid or liquid the outer parts would rotate faster than the inner.

Since the rings of Saturn consist of myriads of tiny bodies how did they get there in the first place? A likely explanation is that they represent the remains of what was once a satellite. Calculations show that when a satellite gets within a certain critical distance of its primary or parent planet it can get completely broken up by powerful tide-raising forces. The bits and pieces gradually get scattered around the primary and eventually form a complete ring or system of rings. Those of Saturn now have an overall width of over 56,000 miles. But since they disappear for a short time when their plane passes through the earth their thickness cannot be more than 60 to 70 miles. They last disappeared in 1950 and will do so again in 1966.

Saturn has nine satellites. Titan, the one discovered by Huygens (p. 39) has a diameter of 3,000 miles and is easily the largest. Phoebe, the smallest and the faintest, has a diameter of about 100 miles. It was discovered by W. H. Pickering in 1898 and moves with retrograde motion at a distance from Saturn of about 8 million miles. So perhaps it is not a genuine satellite but a captured body like some (or maybe all) of the small satellites of Jupiter.

Both Saturn and Jupiter send out bursts of radio energy. Those from Jupiter are by far the stronger but they do not come from the Great Red Spot, nor it seems, from any particular feature or features on the cloud belts. Radio astronomers are therefore investigating the possibility that the bursts are associated with activity on the sun which somehow upsets the planet's magnetic field.

CHAPTER 8

Additions to the Solar System

UNTIL 1781 there were only six known planets. In that year William Herschel, using a small reflecting telescope, discovered a seventh, Uranus. At first he thought it was a comet, for it had a slightly nebulous or misty appearance and slowly changed its place among the stars. The mathematician Lexell calculated its orbit, only to find that it was a planet revolving at roughly twice the distance of Saturn from the sun. The discovery made quite a stir, the more so since Herschel was then almost unknown in the world of science. Everyone had got so used to the idea of six planets that it seemed almost impossible that there should be a seventh. Later, it was found that several astronomers before Herschel had seen Uranus, but all had mistaken it for a star.

Strictly speaking Uranus is a naked-eye planet, although it never gets brighter than the faintest star seen with the unaided eye on a fairly clear night. Its average distance from the sun is 1,782 million miles (an easy number to remember in view of the date of its discovery) and it makes one revolution in just over 84 years. The fact that it shows a definite disc in a telescope of moderate power means that it is a giant planet. The average apparent diameter corresponds to a real equatorial diameter of about 29,300 miles, but this is uncertain owing to the smallness of the angle and the planet's slightly hazy appearance. There is a slight polar flattening of about $\frac{1}{14}$, equal to that of Jupiter and indicating that the period of rotation is fairly short. This is supported by spectrographic observations which suggest a time of 10 hr. 49 min. Further support is given by the low average density of 1·70 which also means that Uranus, despite its great size, has only 14·5 times the mass of the earth.

Spectrograms of Uranus show intense absorption bands due to methane but those of ammonia are quite weak. This fits in with the radiometric estimate of a temperature of less than −185° C. Most of the ammonia must exist in a frozen state. It is highly probable that the physical conditions on Uranus are similar to those on Jupiter and Saturn.

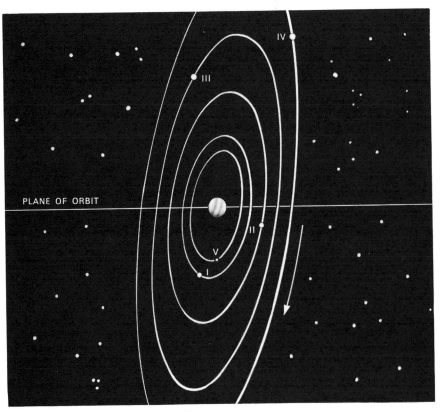

Uranus and the orbits of its satellites.

Uranus has five satellites. William Herschel discovered the two brightest, Titania and Oberon, while two more, Ariel and Umbriel, were detected by William Lassell in 1851. All four have diameters between 750 and 1,750 miles, but these estimates are uncertain since they are based on brightness alone. Miranda, the fifth and by far the faintest, was discovered as recently as 1945 by G. P. Kuiper with the 82-inch reflector of the McDonald Observatory, Texas. Its diameter is probably only about 200 miles.

The curious thing about all five satellites is that the planes of their orbits are tilted almost at right angles to the plane of the planet's orbit. Faint bands or belts sometimes seen on the disc show that this is also true of the plane of the planet's equator. Its polar axis therefore lies close to the plane of its orbit and can sometimes (as in 1903 and 1945) point directly towards the earth. The satellite orbits are then seen almost in plan as nearly perfect circles. Twenty-one years later (as in 1924 and 1966) the axis is roughly at right angles to the line joining the earth and the planet. The orbits of the satellites are then seen edgewise and appear as straight lines. Actually the whole system rotates in a retrograde sense, that is, backwards with respect to its direction of revolution about the sun. It is just as if it once rotated in a direct sense and then somehow got turned upside down.

January 1, 1801, saw another addition to the solar system in the form of a tiny planet named Ceres. Its discoverer, the Italian astronomer Piazzi, was checking star positions in a part of the zodiac at the time and came across it quite by chance. In the telescope it looked like a point of light but the way it moved showed that it was no star, nor even a comet, but another planet. Calculations based on further observations showed that it had a period of about 4·6 years. Its orbit therefore lay between those of Mars and Jupiter. This was not unexpected, for Kepler and others had commented on the large gap between the orbits of these planets. If the radius of the earth's orbit is made equal to 1, the average radii of the orbits of Mars and Jupiter are 1·5 and 5·2 respectively. The orbit of Ceres, with a radius of 2·8, seemed to 'fill the gap' quite well. But at this distance a sizeable planet would have shown a reasonably large disc. Ceres, on the contrary, looked small even under high magnifications. So small, in fact, as to indicate that its diameter was only a few hundred miles.

Micrometric measurements show that the apparent diameter of Ceres at its closest is only about $0''\cdot 3$ of arc. An angle of this small size is extremely difficult to measure even with the best of micrometers, and any small error can affect the result by several miles. According to modern estimates the diameter is between 420 and 500 miles. Ceres is therefore a *minor planet* or *planetoid*, although its star-like appearance has also earned it the name *asteroid*.

The discovery of Ceres encouraged a number of German astronomers to organise a hunt for more planetoids. Success soon came their way. Olbers in 1802 discovered a second, named Pallas, Harding in 1804 picked up a third, Juno, and in 1807 Olbers discovered yet a fourth, Vesta. Their diameters are roughly 300, 120 and 240 miles respectively and they all move in orbits between those of Mars and Jupiter. No more were discovered until 1845 when Hencke found Astraea and, by so doing, gained a pension of 1,200 gold marks from the King of Prussia.

In 1891, when the number of known planetoids had reached 323, Max Wolf of Heidelberg began to use photography for their discovery. This had the advantage of saving valuable time and effort at the telescope. The stars recorded their images on photographic plates which could then be studied at leisure on cloudy nights. One method, now widely used, consists of photographing a part of the sky with an exposure time of several hours. Throughout the exposure the telescope must be kept directed towards the same stars by means of a mechanical or electric drive, otherwise each star will move across the plate and produce a trail of light. A planetoid (or for that matter, a faint comet) then gives itself away by making a short trail.

Wolf and his colleagues have been likened to deep sea fishermen who changed from lines to nets. As plate after plate was exposed and developed the catch of planetoids went up by leaps and bounds. Wolf himself discovered over a hundred and by 1903 the total had passed the five hundred mark. Over 1,600 have now had their orbits determined, but photographic plates show many more whose orbits are not accurately known. Quite often they turn up on photographs taken for quite different reasons and in this way can be great nuisances. Many thousands probably exist but have escaped detection owing to their small size and hence extreme faintness.

Of all the planetoids only Ceres, Pallas, Juno and Vesta are large enough to be measured with a micrometer. The sizes of the rest have to be estimated from their brightness alone. The majority are thought to be no larger than lumps of rock a few yards across. There is no reason to think that they are even roughly spherical. Instead, several change in brightness in a fairly rhythmic way so as to suggest that they are cigar-shaped bodies in a state of rotation. Curiously enough the brightest is Vesta with a diameter roughly one half that of Ceres. Since both objects are almost equally distant from the sun this means that Vesta is the better reflector of the two. While Ceres reflects only about 3 per cent of the sun's. light, Vesta must reflect at least 25 per cent. It may therefore be a metallic object, have an ice-covered surface, or even be made up almost wholly of frozen gases.

Most planetoids move between the orbits of Mars and Jupiter and keep to the general plane of the solar system. Some, however, have orbits which carry them beyond Jupiter while others are taken well within the orbit of the earth. Those that pass fairly close to the earth are useful for determining the sun's parallax.

The most famous planetoid in this connection is Eros, a cigar-shaped body some 15 to 20 miles long, discovered in 1898 by the German astronomer Witt. Calculations based on his observations showed that it would come within 30 million miles of the earth in 1900-1. Photographs were then taken at a number of widely separated observatories and used to determine the planetoid's parallax. Since its period was known (1·76 years), application of Kepler's third law enabled the parallax to be converted into a solar one of $8''\cdot807$ of arc. In 1930-1 Eros made another close visit, this time approaching to only 16 million miles, and 24 observatories collaborated in taking the necessary photographs. The analysis of the results, made by Sir Harold Spencer Jones, the late Astronomer Royal, and not completed until 1941, gave $8''\cdot790$ of arc for the sun's parallax, corresponding to a distance of almost exactly 93 million miles from the earth.

Another planetoid that comes close to the earth is Apollo, discovered in 1932 by K. Reinmuth of Heidelberg. When its orbit was calculated it was found to spend part of its time inside the orbit of Venus and to have approached the earth as close as 2 million miles. Even closer approaches were made by Adonis (1 million miles) and Hermes (5,000 miles) two tiny bodies discovered in 1936 and 1937 respectively. Unfortunately, all three were observed for so short a time that their orbits could be only roughly determined. So unless they pass near the earth on some future trip they may never be seen again. So far, the planetoid with the most unusual orbit is Icarus, discovered in 1949 by Walter Baade of the Mount Wilson and Palomar Observatories. It is about a mile in diameter and sweeps around the sun in a period of only 409 days. Its aphelion sees it well beyond the orbit of Mars, yet at perihelion it passes within 18 million miles of the sun, or about halfway between Mercury and the sun.

Some years after its discovery Uranus was found to be deviating from its calculated orbit. Its observed position did not agree with the predicted one, even after allowing for the gravitational influences of Jupiter and Saturn. One explanation was that a disturbing body in the form of an unseen planet existed beyond Uranus. If so, what was its mass, distance and orbit, and what were the chances of finding it? Only two mathematicians, J. Couch Adams of Cambridge and J. J. Leverrier of Paris dared to tackle so immense a problem. Neither one knew what the other was doing, yet their results agreed remarkably well. Adams produced his in 1845, giving the mass, orbit and probable position of the unseen planet, but he had to wait over a year before anyone would search for it. Meanwhile, Leverrier published his results and sent a predicted position to Galle, director of the Berlin Observatory. Galle received the information on September 23, 1846, and tracked down the planet a few hours later. Not only was the object near the place which Leverrier had predicted but, as it turned out later, it was also near that predicted by Adams.

Neptune, as the new planet was called, takes so long to travel once round the sun that it has not yet made one complete circuit since it was discovered. The period of revolution is 164·8 years, or nearly twice that of Uranus. Even so, the orbital velocity is 3·4 miles a second compared with 4·2 miles a second for Uranus. The orbit is immense, the average distance from the sun being 2,792 million miles, or roughly 30 times the corresponding distance of the earth. Yet despite its great distance it can be seen in a small telescope or pair of binoculars and shows a definite disc in larger instruments.

Micrometric measurements indicate that the actual diameter is 27,800 miles, or less than that of Uranus. The smallness of the disc (about 2" at the most) makes the value uncertain, but it now appears that Neptune is smaller than Uranus and not larger as was previously thought. The disc shows no definite markings or belts. Strong absorption bands in the red part of the spectrum are said to give the planet a greenish colour. The bands are due to methane which, despite the low (calculated) temperature of about $-220°C.$, could still exist as a gas. Any ammonia would probably all be frozen. The rotation period, estimated by means of the Doppler-Fizeau effect, is between 14 and 16 hours.

Neptune has two known satellites: Triton, discovered by Lassell in 1846 and Nereid, first seen by Kuiper on photographs taken in 1949. The motions of these satellites, together with the perturbing influence of Neptune on Uranus, show that Neptune has a mass 17·5 times that of the earth and a fairly high average density of 2·26. Triton has a diameter of about 2,300 miles and is bright enough to be seen in a telescope of moderate size. It moves with a retrograde motion at an average distance of just

ADDITIONS TO THE SOLAR SYSTEM

over 23,000 miles from Neptune. Nereid, so faint that it can be seen only on photographs taken with giant telescopes, cannot be much more than 200 miles across. Its motion is direct but its orbit, the most eccentric of all satellite orbits, carries it to within 1 million miles and beyond 6 million miles of Neptune.

Disappointment awaited those who thought that the influence of Neptune would account for all the remaining deviations in the motion of Uranus. It gradually became obvious that some other unseen planet was also having an effect. Two American astronomers, P. Lowell and W. H. Pickering, therefore tried to find the mass, orbit and probable position of the hypothetical ninth planet. Several searches were made, but without success. Then in 1928 Pickering, not to be outdone, made fresh calculations. A further search was made, this time with the 13-inch photographic

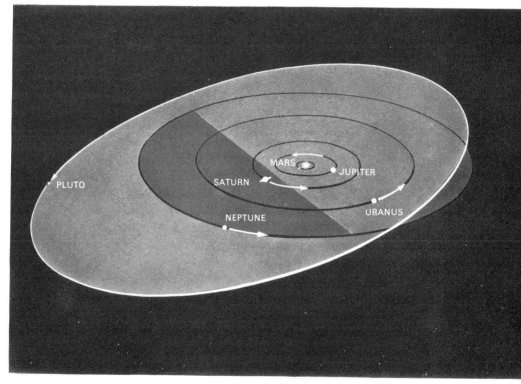

Orbit of Pluto.

refractor of the Lowell Observatory, and on January 23, 1930, Clyde W. Tombaugh tracked the planet down. It moved so slowly that nearly two months passed before the Lowell astronomers could be sure that it was indeed a planet beyond Neptune. The discovery was announced on March 13, 1930, and the planet received the name Pluto.

Pluto is an extraordinary planet. In the first place its orbit is even more eccentric than that of Mercury. Its mean distance from the sun is 3,649 million miles, but at perihelion it is actually closer to the sun than Neptune. At present it is on its way towards perihelion and between 1969 and 2009 will be closer to the sun than Neptune. It reaches perihelion in 1989, but since its period is 248 years, will not get there again until 2237. The orbit is unusual in another way. Its inclination to the plane of the earth's orbit (and therefore to the general plane of the solar system) is just over 17°. This means that, unlike the other planets, Pluto can wander outside the zodiac. Also, that it cannot possibly collide with Neptune.

Pluto turned out to be much fainter than expected. To produce the observed perturbing effects on Uranus and Neptune it had to be almost as massive as the earth. Yet measurements in 1950 by Humason and Kuiper with the 200-inch Hale telescope indicated a diameter of only 3,600 miles. On this basis the density would be about 12 times that of the earth—an improbable result to say the least. More recent measurements put the diameter at less than 4,900 miles, but the density still comes to about 6 times that of the earth. An ingenious suggestion is that large or numerous 'seas' of liquid or frozen methane give Pluto's surface a polished appearance and make it look smaller than it really is. The distance and period of a satellite would, of course, give a direct answer to the problem of the planet's mass, but so far none has been detected. Another oddity is the length of the period of rotation. Small but rhythmic light variations detected in 1955 indicated an unusually long period of about 6 days 9 hours 17 minutes.

CHAPTER 9

Comets and Meteors

MENTION of the unusual reminds us of comets. As far as naked-eye observation is concerned they can arrive suddenly as if coming from nowhere, move steadily across the background of stars for a few days or weeks and then disappear. Aristotle thought that they were no more than small regions of fiery air high up in the atmosphere. That like shooting stars and rainbows, they belonged to the region of change centred on the earth and enclosed by the shell or sphere of the moon. This view became generally accepted and lasted right up to the sixteenth century, largely because it fitted in with the Aristotelian idea of the unchanging heavens.

In 1577 Tycho Brahe observed a bright comet and attempted to measure its parallax. If Aristotle were right its parallax should be larger than that of the more distant moon. Tycho found just the reverse—the parallax was less than 20'. He also saw that the comet's tail pointed away from Venus and not away from the earth as required by Aristotle. As a result he broke with tradition and gave the comet a circular orbit outside that of Venus.

Halley took the next step. Using a method devised by Newton and based on the inverse square law of gravitation, Halley calculated the orbits of 24 bright comets and found that those of 1531, 1607 and 1682 were strikingly similar. He therefore concluded that the comets of those years were one and the same object which moved around the sun with a period of about 76 years. He also predicted that it would return about 1758. And return it did, although somewhat delayed, as he had foreseen, by the attraction of Jupiter. It was first seen on Christmas Day, 1758, but not by Halley. He died in 1742 and therefore did not live to see this great triumph of the Newtonian theory. Through his labours one comet at least became a permanent member of the solar system. This being so it was reasonable to suppose that most, if not all, other comets were also members.

Historical research shows that Comet Halley has been seen over a

Halley's Comet, 1910.

period of at least 2,000 years. It certainly appeared in 240 B.C. and also in A.D. 1066, just before the Battle of Hastings. It last appeared in 1910 and all being well should return in 1986-7. Its highly eccentric orbit puts its perihelion closer to the sun than the earth and its aphelion well beyond the orbit of Neptune. It therefore has every chance of coming under the perturbing influence of one of the outer planets. For this reason its period has been as long as 79·3 years and as short as 74·5 years and the date of its next return cannot as yet be accurately predicted.

The main feature of Comet Halley is its long, bright tail. Many other (but certainly not all) comets share this distinction, but no two tails are alike in length, brightness and structure. Long-tailed comets tend to be rather rare and their unexpected appearances and unusual shapes used to strike terror in the minds of ordinary folk. In early Christian times the tail was seen as a sword or pillar of fire put in the sky by an angry God to warn the wicked of His coming wrath. As the tail grew it was supposed to herald widespread disaster, and perhaps, the end of the world. Yet surprisingly enough Comet Halley and many other prominent comets show no tails at all when they are at great distances from the sun. They then look like faint misty stars and are extremely difficult to detect. They develop tails only when they come near to the sun. The tails always point away from the sun but may take up slightly different directions owing to their great speeds through space. This suggests that a tail is formed by the sun's heating action on the *nucleus* or main body of a comet. Also, that the material so formed is then blown away from the nucleus by some sort of solar pressure or 'wind'. A comet therefore loses mass every time it passes near the sun and cannot be classed as a permanent member of the solar system.

It is now fairly well established that these interpretations are correct. Although a tail can extend many million miles it is such an empty affair that faint stars can be seen through it without being perceptibly dimmed. Its material must therefore be finely divided and highly rarefied. Spectrograms confirm this by showing bright lines and bands on a faint continuous background. The lines and bands are due to compounds of carbon, hydrogen, nitrogen and oxygen in the form of molecules. They also indicate that the gas is very thin and is being made to glow by the energy received from the sun. They develop as the comet approaches the sun but those due to cyanogen, a compound of carbon and nitrogen, are present practically all the time. The faint background spectrum comes from sunlight reflected by widely scattered particles of dust. In some cases rapid movements in the tail cause it to change its appearance almost minute by minute and show that it is in a state of great turbulence. Comet Morehouse 1908, and Comet Mrkos 1957, both possessed exceptionally active tails.

Until recently it was thought that cometary tails were driven out by sunlight. Laboratory experiments showed that light exerted an extremely small pressure on solid objects placed in its path. Calculations indicated that if the light were intense and the objects had diameters about equal to the average wavelength of light they could be pushed along. It is now known, however, that a more powerful agent is at work in the form of a stream of electrified hydrogen gas which comes from the sun. The stream leaves the sun's atmosphere in the form of an expanding shell which moves faster as the distance from the sun increases. In the region of the earth the speed is thought to be about 250 miles a second. As yet, the study of this 'solar wind', as it is called, is still in its infancy, but it is now pretty clear that its swift flow, aided by the pressure of light, produces the tails of comets.

To the unaided eye, and also through a large telescope, the nucleus of a comet looks like a star. Surrounding it is a mist or haze referred to as the *coma*. Nucleus and coma together form the *head* of a comet. The curious thing about the nucleus is that it is small and has little mass. Just how small it must be was shown on May 18, 1910, when Halley's comet passed directly between the earth and the sun. Despite careful searches over the sun's face not a trace of the comet was seen. The small mass means that a comet can be easily perturbed by a planet, especially when that planet is massive like Jupiter.

From what little evidence we have it seems probable that the nucleus is made up of objects each less than a mile in diameter. They may be giant-sized chunks, tiny particles like grains of sand, or a mixture of both. The English astronomer Raymond A. Lyttleton suggests that a comet resembles a giant swarm of gnats, the swarm being densest in the nucleus and thinnest in the tail. According to the American astronomer Fred Whipple, the nucleus consists of a spongy mass of lumps of stone and metal held together by ices. When the comet is in the depths of interplanetary space the ices are in the form of frozen gases like methane, carbon dioxide and ammonia. But as it nears the sun the ices evaporate to form a coma and then a tail.

The idea that a comet consists largely if not completely of a swarm of particles is supported by an interesting association between comets and meteors. Every November, shooting stars known as the Leonids appear to travel or radiate from a point in the constellation of Leo. They have been doing this for centuries but received hardly any attention at all until the night of November 12, 1833. Instead of giving the usual 'shower' they almost covered the sky with their shining tracks. Their numbers were quite beyond counting although one estimate put them at over 26,000 an hour. They returned to normal in the following years but on the night of November 13, 1866, gave another brilliant display. Vast numbers fell throughout the night but astronomers who had seen the 'storm' of 1833 thought that the second event was less impressive.

By this time H. A. Newton in America and G. Schiaparelli in Italy had found an explanation of both showers and storms of Leonids. A stream of meteors, they suggested, was spread around the sun in a track or orbit. Every November the earth passed through the track and swept some of the meteors into its atmosphere. The fact that the Leonids appeared to radiate from a single point (the *radiant*) showed that they travelled in parallel paths in space. A similar effect is seen with the lines of a long straight railway track. Although the lines run parallel to each other they appear to diverge from a point in the far distance. The fact that storms of Leonids occurred in 1833 and 1866 meant that the meteors were much more densely packed in one part of the track than elsewhere. Schiaparelli, and independently Leverrier, also found that the orbit of the Leonids resembled that of Comet Tempel-Tuttle 1866, a faint comet with a period of just over 33 years.

A further link in the comet-meteor chain was forged in 1872. At its return in 1846, a periodic comet named Biela startled astronomers by appearing double. The head actually broke into two parts, each with its own nucleus and tail. The two comets appeared again in 1852, but the separation between them had increased to nearly 2 million miles. They then disappeared, but in 1872, when the comet was due to reappear, a splendid shower of meteors occurred from a radiant in Andromeda. Calculations showed that the orbit of the meteor swarm coincided with that of Biela's comet, or what was left of the comet. Another fine display occurred in 1855, but since then the Bielids or Andromedes have given only small showers. Several other comets have been observed to break up, but without providing any meteor showers or clues as to how the divisions took place.

Very few meteor streams have given spectacular showers. The Leonids proved disappointing in 1899 and 1932. Saturn and Jupiter had apparently perturbed their orbit and the earth passed through only the fringes of the main swarm. In this century the best displays have been given by the

October Draconids, or Giacobinids, of 1933 and 1946. Their radiant in Draco was where it would have been had they been moving in the orbit of Comet Giacobini-Zinner. At the heights of both these brief but intense showers, meteors appeared at the rate of about 350 a minute. Two annual showers, the Eta Aquarids in May and the Orionids in October, are associated with Comet Halley, while the Perseids in August appear to be linked with Comet Swift-Tuttle 1862.

What is the physical connection, if any, between a meteor stream and its associated comet? As early as 1867 D. Kirkwood in America suggested that meteor streams were the remains of comets. That as a comet slowly wasted away its material became spread along its orbit. Eventually, when the comet was no more, its debris in the form of meteoric particles and dust was scattered more or less evenly all along the orbit. How correct this is we cannot say. Much more evidence is necessary before a decision can be reached. Several meteor streams have still to be associated with comets and there must be many more waiting to be discovered. We get to know about a stream only when the earth passes through it, and since the earth is only a tiny speck in the immensity of the solar system it can easily miss a stream completely.

Comets are now being discovered at the rate of five to six a year. Most of them are very distant and never become visible to the unaided eye. The 800 or so which appear in catalogues probably represent only a tiny fraction of the total number. Perhaps, as Kepler once remarked, they are as numerous as fishes in the sea. They not only populate the solar system but extend its frontiers far beyond the orbit of Pluto. A recent object, Comet Mrkos 1957, is estimated to travel 28 times farther from the sun than Pluto and to have a period of 13,000 years. Yet this is a fairly frequent visitor to the sun compared with others thought to stay away for hundreds of thousands of years and to travel almost half-way towards the nearest stars. Once a comet is near the solar system, however, planetary perturbations can easily swing it into a much smaller orbit or perhaps send it away forever. At the other end of the scale there are short-period comets which orbit the sun in only a few years. One of these is Comet Encke, a well-observed but fairly faint object with a period of 3·3 years. Another, Comet Wilson-Harrington, has an even shorter period of 2·3 years. Unfortunately, it was lost soon after its discovery and has not been seen since.

COMETS AND METEORS · 93

The solar system has a vast population of meteors in addition to those that make up the orbiting streams. The great majority wander like vagabonds in space, ruled over by the sun but easily influenced by any massive body that happens to pass by. They range in size from microscopic specks to giant chunks weighing many tons. It is estimated that thousands enter the earth's atmosphere every second and that during a year they add about 5 million tons to the earth. If this latter figure is anywhere near correct it is still next to nothing compared to the mass of the earth itself.

According to Whipple, extremely small meteors or micrometeorites get through the atmosphere without damage. Those slightly larger than this but smaller than a pea get vapourised by the heat of friction with the air. They enter the atmosphere at speeds of 7 to 45 miles a second, travelling slow when they overtake the earth and fast when they meet it head-on. Although the air through which they pass is very thin, the heat of

Earth passing through a meteor stream.

friction due to the great speed is sufficient to raise their temperatures several thousand degrees. They are therefore vapourised at heights above 60 miles, but in the process cast off drops of molten metal. These then cool, solidify, and fall to join the micrometeorites on the ground and the beds of seas and oceans. Finally, meteorites more than about 20 pounds in weight reach the ground in one whole piece but lose much material in the process. The largest discovered so far weighs about 60 tons and still lies where it fell near Grootfontein in South West Africa. The second largest weighs 34 tons and is on show in the American Museum, Hayden Planetarium, New York.

No large meteorite has as yet fallen in a town or city. Some idea of the damage it could cause was shown in June, 1908, when an intensely bright meteor sped over central Siberia and crashed in the remote Tungus forest in northern Siberia. When it fell it flattened trees for miles around, sent a great cloud of dust into the atmosphere, and shook the ground so violently that an earth tremor was recorded all over Europe. When a Soviet expedition visited the area some years later it found a number of small craters in marshy ground but no traces of meteoritic material. A similar explosion took place in February, 1947, in the Sikhote-Alin Mountains north of Vladivostok. The shock wave felled trees over a large area but this time Soviet expeditions were able to collect about 23 tons of meteoritic material. Once again the craters were small but numerous. The official Soviet view is that both explosions took place several hundred feet above the ground. The Tungus object, it is suggested, was a cometoid, or mass of frozen gases similar to that believed to form the nucleus of a comet.

Meteorites are the only bodies in space about which we have direct knowledge. Laboratory tests show that they can be grouped into three main classes or types: *siderites*, composed mainly of nickel-iron, *aerolites*, stones with perhaps a little nickel-iron, and *siderolites*, bodies which contain about equal amounts of stone and iron. Those of any one class have different structures and often contain different amounts of other substances such as copper, zinc, cobalt and glass. Some metallic meteorites have also been found to contain free carbon in the form of graphite and tiny but clear crystals of diamond.

CHAPTER 10

The Sun

ONE way of finding the distance of a body in space from the earth would be to flash an intense beam of light at it and find out how long the light takes to get there and back. Since the speed of light is known the distance would follow by simple arithmetic. This may sound fantastic, but a source has recently been developed which gives an intensely brilliant and parallel beam of light. With this source, known as the *laser*, it is possible to send a light signal to the moon and, $2\frac{1}{2}$ seconds later, to pick up part of the reflected beam. Radio waves travel at the same speed as light, and by using giant aerials, narrow beams of intense short-wave radio waves can be sent to the moon and planets. The technique of using beams of this kind is known as *radar* and was developed during World War II for locating and tracking aircraft. In 1946 Bay of Hungary and technicians of the U.S. Army Signal Corps detected the return of a radar signal reflected by the moon. Twelve years later workers at the Lincoln Laboratory of the Massachusetts Institute of Technology used a radar system 10 million times more powerful and got a reflected signal from Venus. In the following year, 1959, a radio telescope at Stanford University picked up the first radar echo from the sun.

In view of the last achievement it might be thought that astronomers now have a direct way of measuring the sun's distance. Unfortunately the sun is itself a powerful source of radio waves and almost completely drowns out ordinary radar echoes. It also acts as a 'soft' reflector, that is, it does not present a solid surface. Its extensive atmosphere absorbs and scatters most of the energy of a radar pulse and reflects the part that does remain long before it can reach the sun's main body. The sun's distance is therefore best obtained by using a 'hard' body like Venus. This has now been done at a number of stations in the U.S.A., the U.S.S.R., and also at Jodrell Bank, near Manchester. Giant radio telescopes were converted to send intense radar signals at the close approach of Venus late in 1962 and gave fairly uniform results. They reduced the uncertainty of the sun's parallax to one part in 100,000 and made its mean distance 93,499,000 ±300 miles. Radar echoes have since been received from Mercury, Mars and Jupiter, but apparently none has been used to improve upon the result obtained with Venus.

Once the sun's average distance from the earth is known it is a simple matter to determine its size. It is only necessary to measure the sun's apparent diameter. This on the average is 32′ of arc and puts the real diameter at 865,000 miles, or a little over a hundred times the diameter of the earth. Actually, the apparent diameter varies from day to day as the earth revolves about the sun in its elliptical orbit, but it always falls between the limits 32′·6 of arc (when the earth is at perihelion) and 31′·5 of arc (when it is at aphelion). The sun's mass is determined by considering the motion of the earth. The very fact that the earth travels around the sun means that it is continually falling towards the sun. If it failed to do this it would fly off along a tangent to its orbit like a stone suddenly released from a sling. Calculation shows that in one second (or in travelling $18\frac{1}{2}$ miles) the earth is drawn towards the sun by a little more than one ninth of an inch. This deflection from what would otherwise be motion in a straight line is a measure of the force with which the sun attracts the earth. This force, by Newton's law of gravitation, is proportional to the product of the masses of the earth and the sun, and decreases as the square of the distance between their centres. The earth's mass has been found by several methods, all based on the law of gravitation. So knowing the attractive force, the earth's mass, and the distance between the earth and the sun, we can calculate the sun's mass. The answer is that the sun is 333,400 times more massive than the earth. From the mass and volume the average density of the sun's material is found to be 1·4, or about $\frac{1}{4}$ the average density of the earth.

The intensely bright visible surface of the sun is called the *photosphere* or 'light-sphere'. To the unaided eye it looks equally bright all over, but the telescope (and especially photographs taken with the telescope) shows that it has a definite darkening towards the edge or limb. This is due to absorption in the sun's atmosphere. Light coming from the limb has to pass a greater distance through the sun's atmosphere than that coming from the centre of the disc. Even the centre of the disc is not uniformly bright. Observations made with high magnifications when the air is steady show that the central parts are speckled or granulated. The

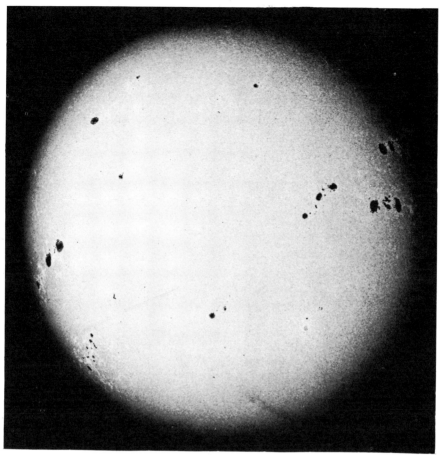

The sun, showing a large number of sunspots and several faculae.

small bright specks which give rise to this appearance are called *granules*. They look small but in fact average about 500 miles in diameter. Irregular in shape, they are separated by slightly darker and therefore cooler spaces. They last for only a few minutes at most and are thought to be the tops of currents of gas which carry the sun's heat from the interior on its last journey into space.

Another interesting feature of the photosphere is large bright patches known as *faculae*. They are best seen near the limb where they stand out in comparison to the darker surrounding photosphere. Usually they are found near sunspots but they sometimes appear when spots are altogether absent. They do not hide the granules and are thought to rise like intensely bright hills and mountains above the general level of the photosphere. Even so they have never been seen standing out beyond the limb.

The most striking events on the photosphere are undoubtedly the dark spots observed by Galileo and noticed with the naked eye long before his time. Galileo was the first to realise that the spots belonged to the sun. He noticed that they appeared to move steadily from east to west across the sun's face in about 13 to 14 days and concluded that the sun rotated in a period of about 27 days. This period, however, is measured from the earth which is travelling in the same direction as the sun is turning. When allowance is made for this the sun is found to rotate once in 25 days at its equator. The paths of the spots are straight lines in June and December. This means that the earth is then in the plane of the sun's equator. In March they are curved to the north (when the sun's south pole is tipped towards the earth) and in September they are curved to the south (when the sun's north pole is tipped towards the earth). Measurements based on these changes show that the sun's equator is tilted 7° 15' to the plane of the earth's orbit.

In 1860, Richard Carrington, an English amateur astronomer, drew attention to the fact that the sun's period of rotation was shortest at the equator. Away from the equator it increased with increasing latitude to reach 27·7 days at latitudes 45° N. and S. Spots seldom appear in latitudes higher than 45°, but studies of the Doppler-Fizeau effect based on spectrographic observations show that the period continues to lengthen right up to the sun's poles. This can mean only one thing—the photosphere is not a solid surface. Indeed, we know from other considerations that the sun is composed of gases right through to its very centre. The gases deep down are extremely hot and under tremendous pressures: even in the outer levels they are dense enough to give the sun a solid-looking surface. Dense enough, also, to emit light of all wavelengths and so produce a continuous spectrum. As we saw earlier, the dark lines in the sun's spectrum are produced by cooler absorbing gases *above* the photosphere.

Sunspots vary in size from tiny *pores* about 500 miles in width to giants some 50,000 miles across. They occur singly and in groups and first appear as small dark patches among the granules. They grow rapidly, the small ones to last less than a day, the large ones to survive for several days and, maybe, weeks. The largest group ever seen appeared in April,

1947, and had an area of over 6,200 million square miles. The remarkable thing, as Heinrich Schwabe of Dessau discovered in 1843, is that sunspots vary in number in a fairly regular way. Using only a small telescope, Schwabe kept close watch on the sun for 18 years. As the years went by he realised that the number of groups rose and fell with a general period of 10 years. Since then further studies along similar lines have shown that the period varies from 7 to 17 years and has an average value of 11·1 years.

The change in sunspot numbers during each period is known as the *sunspot cycle*. At sunspot minimum, or the beginning of a cycle, the sun is fairly free from spots for weeks and perhaps months on end. Spots then appear in fairly high latitudes (about 30° to 35° N. and S.) and increase in number to reach a maximum after about four and a half years. The numbers then decrease and get to a minimum again about six and a half years after the maximum. All the time the places where the new spots appear gradually approach the sun's equator. So as the cycle nears its end the spots always appear near the equator. By this time additional spots may begin to put in an appearance in fairly high latitudes. They belong to the next cycle, but like the remaining spots of the previous one, are usually few in number.

Sunspots generally consist of two distinct parts—a dark central region or *umbra* surrounded by a lighter border known as the *penumbra*. Near the sun's limb they become so highly foreshortened as to appear like narrow dark strips. Most of them then look like hollows in the photosphere. Spectrograms show that they have temperatures of about 4,000° C., or some 2,000° C. less than that of the surrounding photosphere. They are therefore quite bright but appear dark by comparison with the brighter background of the photosphere.

In the late 1880's the American astronomer G. E. Hale invented the *spectroheliograph*, a special instrument which enabled him to photograph the sun in the light of a selected wavelength. In principle the slit of a powerful spectrograph was made to move slowly over the sun's disc while a second slit, set on a particular absorption line in the spectrum, moved at the same rate over a photographic plate. Photographs taken in this way using a strong line in the red due to hydrogen showed that some spots were surrounded by whorls of hydrogen gas. Later, Hale discovered that the lines in the spectra of spots were split into a number of distinct component lines. This appearance, called the *Zeeman effect,* occurs whenever

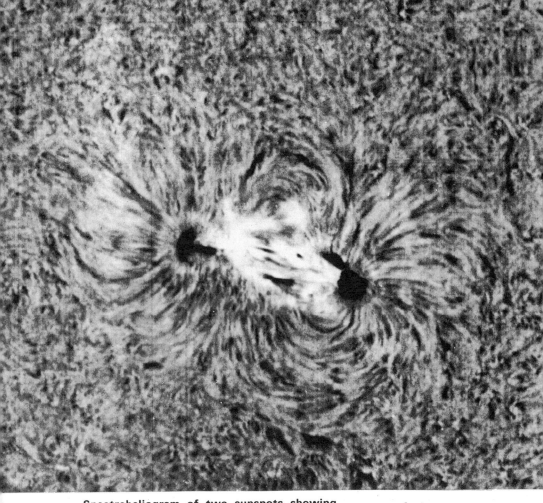

Spectroheliogram of two sunspots showing distinct whorls of hydrogen gas.

light passes through a strong magnetic field. Its presence in sunspot spectra therefore means that each spot is the centre of a magnetic field.

Hale and his colleagues found that when two spots occurred as a pair their magnetic polarities were of opposite types. If the magnetism of one spot was 'north' the other was 'south', and vice versa. Also, that the first or leader spot of any pair in one hemisphere had a polarity opposite to that of a leader spot in the other hemisphere. For instance, if the leader in the northern hemisphere had north polarity its counterpart in the southern had south polarity. Yet, strange to tell, in the next sunspot cycle the polarities switched over. Leader spots in the northern had south

THE SUN · 101

polarity and those in the southern had north. The reversal in polarity always took place with the start of each new sunspot cycle, but the whole process of magnetic change took twice as long as the average sunspot cycle.

What, then, are sunspots? Several theories about them have been put forward but none is really satisfactory. This is not surprising in view of their remarkable behaviour. At present they are considered to be fairly quiet regions in an otherwise stormy photosphere. They behave as if they were giant electromagnets, obtaining their strong magnetic fields from currents of electricity in the form of circulating streams of electrons. Hale thought they were giant whirlpools of gas rather like cyclones in the earth's atmosphere, but this idea is no longer generally accepted.

During a total eclipse of the sun the dark body of the moon is seen to have a narrow rosy-red fringe and to be surrounded by an extensive halo of pearly-white light. The fringe is the sun's *chromosphere*, a region of extremely rarefied gases which extends some 10,000 miles above the photosphere. The halo is the *corona*, the even thinner 'outer' atmosphere of the sun. So, strictly speaking, the moon can totally eclipse only the photosphere. The rest of the sun remains uncovered.

In the lower levels of the chromosphere are gaseous elements which, by absorbing some of the light from the hotter photosphere directly below, give rise to the dark lines of the solar spectrum. In the upper levels, only

Small-scale loop prominences.

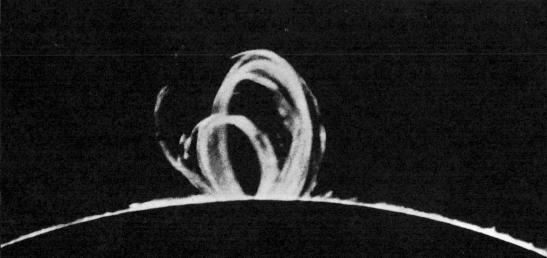

lighter elements such as hydrogen, calcium and helium seem to exist. All levels are the playground of *prominences*, enormous flame-like masses of hot gas often seen in profile on the sun's limb during total eclipses. They can, however, be photographed in broad daylight both on the limb and over the entire disc by means of the spectroheliograph and special optical filters known as *monochromators*. The latter let through light of only one particular wavelength and are usually designed for the deep red light of hydrogen. They also enable a series of photographs to be taken at about two-minute intervals which can then be run off as a cinematograph film. Motion pictures of this kind provide a great deal of information, especially with regard to prominences which happen to appear on the limb. They show, among other things, that some prominences in the form of jets and streamers follow looped paths as if controlled by lines of magnetic force. Many first appear high up in the chromosphere and then pour down to the surface like majestic fountains of fire. The effect is just as if the film were being run backwards. Some are shot outwards with explosive violence to reach heights of hundreds of thousands of miles. Others, much less active, form pyramids, arches and patterns like the outlines of trees. There are, in fact, several different classes of prominence but they all witness to a scene of violent and constant change in an atmosphere which is at once vast in extent and almost overwhelming in its complexity.

Prominences and the upper chromosphere can also be studied with the *coronagraph*, an instrument invented by the French astronomer B. Lyot. In this the image of the photosphere formed by the object-glass of a telescope is artificially eclipsed by a round black disc. The lenses have to be free from dust, scratches and other blemishes which would otherwise scatter the sun's light and produce a glare brighter than the glow of the chromosphere and corona. Dust particles in the earth's atmosphere also produce glare, so the coronagraph has always to be used at high-altitude stations like the Pic-du-Midi Observatory in the Pyrenees and the Sacramento Peak Observatory, New Mexico. When these and other precautions are taken the upper chromosphere and *inner corona* (or lower levels of the corona) can be seen direct, have their light dispersed by a spectrograph, or be photographed in the red light of hydrogen with a monochromator.

The coronagraph shows that the top of the chromosphere bristles with fuzzy spikes known as *spicules*. Each spicule is a geyser-like column of shining hydrogen gas some 500 miles in diameter. The gas shoots up at 20 miles a second, becomes invisible about 10,000 miles above the photosphere, and is thought to inject protons (p. 61) into the corona. The protons then presumably stream away from the sun to form the solar wind.

The inner corona extends above the photosphere to about one third the diameter of the sun. Its spectrum consists of a few bright lines on a continuous background. The bright lines are produced by ionised atoms, or atoms stripped of some, if not all, of their electrons. The continuous background shows that the inner corona shines largely by light scattered by fast-moving electrons. Other optical studies confirm this, and indicate that the temperature is about one million degrees Centigrade. While this figure may at first seem absurdly high it must be realised that the entire corona is exceedingly thin. So thin, in fact, as to send out hardly any light of its own. A similar effect is seen in the flame of a Bunsen burner. The flame appears bright yellow when the air supply is cut down but gets fainter, more transparent, and hotter as the air supply is increased. The high temperature of the inner corona is also confirmed by observations with radio telescopes. They also show that the corona extends far into space and thereby makes the so-called 'radio sun' much larger than the familiar 'optical sun'. The radio emission, generally referred to as 'radio noise', is found to be more intense than average when active sunspots are about. At times it increases in sudden 'bursts', some of which, as we shall see later, are connected with *solar flares*, or great eruptions in the chromosphere.

During a total eclipse of the sun the corona can be traced over great distances. Its form and brilliance vary with the sunspot cycle. At sunspot maximum it is fairly evenly spread around the sun. At minimum, great curved streamers stretch out in the equatorial regions and short tufts and plumes appear at the poles. Between times the corona has an 'intermediate' form, with only slight equatorial extensions. Just how far it appears to extend depends largely on atmospheric conditions. Far away from the sun it is so faint that it gets drowned in the general background glow of the sky. But at high altitudes, where the sky during a total eclipse looks quite dark, the corona can be traced to distances of over 25 solar diameters. Nor does it end there. Recent observations indicate that it merges with the *zodiacal light*, a faint glow seen only on favourable

The sun's corona during the total eclipse of January, 1908.

occasions shortly after sunset or before sunrise. The zodiacal light usually appears as a cone-shaped whitish haze, broadest and brightest at the horizon, which stretches along the zodiac. Optical studies suggest that it is due mainly to light scattered by dust particles. If so, the dust must entirely surround the sun and be concentrated in the general plane of the solar system. It probably extends right beyond the earth to the most distant planets. One cannot but wonder whether it represents the material left over from a cloud of cosmic dust which, according to modern views, condensed long ago to form the planets and even the sun itself.

One of the most interesting aspects of science is the way it can sometimes relate two or more completely different events. Several instances of this have already been given, but the one we are now going to consider is perhaps the most remarkable of all. It concerns the close connection between sunspots, lights in the sky, and sudden changes in the direction of a compass needle. The lights referred to are *aurorae*, or the northern and southern lights. They are flickering patches and patterns of coloured light in the night sky most often seen in places about 23° from the earth's magnetic poles. Sometimes they look like filmy clouds of red, green or white light and may pass unnoticed unless the sky is very dark. At their best they form bright ribbons and curtains of light which span the sky, and by their vivid colours and changes in shape, form a spectacle of great beauty.

In 1741 Olaf Hiorter of Uppsala noticed that compass needles behaved strangely whenever there were vivid displays of aurorae. Instead of pointing to the magnetic poles the needles swung about violently. There the matter rested until 1852 when the English scientist Edward Sabine found that the frequency of these disturbances kept pace with the sunspot cycle. They were most intense and numerous during the times of sunspot maxima and hardly ever occurred at sunspot minima. Disturbances of this kind are, of course, due to disturbances in the earth's magnetic field. They are known as *magnetic storms* and, as Hiorter had noticed, go hand in hand with the frequency of aurorae.

Do sunspots therefore disturb the earth's magnetic field and thereby cause aurorae? Not really. Sunspots merely indicate the general trend of change in solar activity. Much more important, as far as the earth is concerned, are solar flares. They were first noticed in 1859 by Carrington and G. Hodgson as two sudden brightenings near a group of sunspots. They lasted for about five minutes and then disappeared completely. We now know that only the very largest and brightest flares can be seen in this way. Many more are shown up by monochromators adjusted for the light of the red hydrogen line. Small ones are so numerous that as many as a hundred can be seen in a single day. All of them, large and small, occur near faculae and active spots and are therefore fairly closely linked with the sunspot cycle.

The energy sent out by a really big flare is thought to equal the explosion of 1,000 million megaton atom bombs. The parts that affect the earth arrive in two forms—waves and particles. The waves include an intense burst of ultraviolet and x-ray radiation which reaches the earth 8·30 minutes after the explosion. This disturbs the earth's ionosphere which in turn affects both short- and long-wave radio reception. The particles, electrons and the nuclei of atoms, travel more slowly. Those of very low energy and speed take between 20 and 40 hours to reach the earth, and then only when the earth lies in the line of fire. They then give rise to aurorae and disturb radio reception. Those of high energy, travelling at a speed approaching that of light, take less than an hour. They consist mainly of protons and are known as *cosmic rays*. Fortunately for us they lose their high energy as they travel through the earth's atmosphere. But in space they could, on hitting the walls of a space vehicle, produce enough x-rays to kill the crew inside. Their accurate prediction will therefore be one of the great problems confronting space travellers of the future.

Thanks to instruments carried by artificial earth satellites and space probes we now know that the earth's magnetic field acts like a huge trap for electrified particles. Most of them, if not all, are thought to come from the sun. Those of low energy spiral around the lines of magnetic force and are guided towards the magnetic poles. They are therefore very numerous in a zone or ring some 23° from the magnetic poles where they

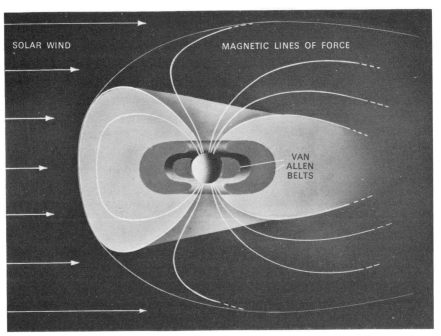

Earth's magnetosphere.

give rise to intense and frequent displays of aurorae. They do this, it is thought, in much the same way as a stream of electrons (that is, an electric current) makes the gas glow in neon and other gas-discharge tubes used in advertising. Parallax observations made from different stations indicate that aurorae occur between 50 and 600 miles above the ground. At these heights the air is so very thin that the passage of streams of electrons and protons could readily make oxygen glow green and nitrogen red.

The region in which electrons and protons are 'trapped' in this way is known as the *magnetosphere*. Its outer parts interact violently with the solar wind, that is, with the hot electrified hydrogen gas that streams away from the corona. Owing to the force of the wind it has a shape like that of a tadpole, the 'tail' being in the plane of the earth's magnetic equator and in line with the sun. When the wind is weak the sunlit side of the magnetosphere is about 40,000 miles from the earth, but when it is strong it can be as close as 25,000 miles. Just how far the magnetosphere extends on the earth's dark side is not known but it may well be several hundred thousand miles. Inside the magnetosphere are two belts or zones in which the energy of the trapped particles is particularly high. These, the *Van Allen belts*, lie above the earth's magnetic equator and disappear completely above the poles. For some reason the intensity of the inner belt varies much more than that of the outer, and at times of high intensity, could be a serious hazard to a spaceman travelling through it. But how these belts are formed and how the particles in them get their high energies is not known. One thing, however, is fairly certain—the particles come mainly from the sun.

CHAPTER 11

A System of Stars

GALILEO'S discovery that the Milky Way was made up of stars set astronomers thinking. Did these stars appear faint because they really were faint? If so, the followers of Aristotle might be right after all in thinking that all stars, bright and faint, were at the same distance from the earth. Or did those in the Milky Way appear faint because they were farther away than the brighter stars? If so, the stars could not possibly be fixed to the surface of a sphere centred on the earth. For a time there seemed to be no way of ever solving the problem. Then in 1717 Halley discovered that the stars *Sirius*, *Aldebaran* and *Arcturus* had definitely moved away from their positions as given in a catalogue by Ptolemy. All three, Halley explained, are bright stars. They are therefore probably fairly near to us as stars go. So if they have any motion of their own they will most likely change their positions with respect to the rest. How right he was! We now know that although the stars appear to be fixed they are all travelling through space at many miles a second. But so far away are even the nearest stars that several thousand years would have to go by before we could, with the unaided eye, notice any changes in their relative positions.

The angle through which a star moves in a given time is called its *proper motion*. *Sirius*, a fairly near star, has a proper motion of $1''\cdot2$ of arc a year. *Aldebaran*, over seven times more distant, has one of only $0''\cdot19$ of arc a year. This means, of course, that proper motions are important clues to distances. It doesn't follow, however, that a star with a proper motion of $1''$ of arc a year is at ten times the distance of one with a proper motion of $10''$ of arc a year, and so on. The proper motion is usually only one component of a star's motion. The other is its motion directly towards or away from us. Fortunately, this second component can easily be measured by means of the Doppler-Fizeau effect. If the star is approaching, the lines in its spectrum are shifted towards the violet end of the spectrum. If it is receding they are shifted towards the red. In both cases the amount of the shift depends on the star's radial velocity, or velocity towards or away from us. Radial velocities are therefore easily and accurately found, but in every case allowance must be made for the velocity of the observer due to the earth's rotation and orbital motion. When this has been done the values obtained are radial velocities relative to the sun. They can then be used either on their own or together with proper motions to provide information about the distances of the stars.

Several astronomers in the eighteenth century hoped that the study of proper motions would reveal the speed and direction of the sun's motion through space. For if the sun were a moving star, stars nearby would show a tendency to drift or stream in a definite direction. Those directly ahead would get wider apart, those behind would get close together, while those alongside would appear to drift in a direction opposite to that of the motion of the sun. At first, since the proper motions of only a few stars were known with reasonable accuracy, no trend of this kind was found. In 1783, however, William Herschel announced that he had obtained evidence of a small but general drift which indicated that the sun was travelling towards a point, the *solar apex*, in the constellation of Hercules. This has since been confirmed, again by studies of proper motions and also by radial velocities, but the precise position in Hercules of the solar apex is still undecided. It is, at any rate, known that the sun (and hence the entire solar system) is moving at about 12 miles a second relative to the bright stars in the sun's neighbourhood.

Herschel also discovered that some stars are not single objects at all but systems of two or more stars which travel around their respective centres of gravity. They are, so to speak, tied together by the invisible bonds of gravitation. When the two components of a double star are associated in this way they are said to form a *binary system*.

The discovery was made in this way: Herschel knew that some stars had a much fainter 'companion' star beside them. It seemed reasonable to conclude that although the two stars appeared to be close together, the brighter one was much closer than the fainter. If so, the earth's motion about the sun would cause the brighter star to appear to travel around the fainter once in a year. But although the method seemed an excellent way of finding stellar parallax, Herschel had no success with it. He did, however, find that several double stars did change their relative positions. And in such a way as to show quite clearly that they were binaries.

Using reflecting telescopes which he made himself, Herschel embarked on a number of explorations of the starry sky. During these he discovered and catalogued several thousand double stars, some of which turned out to be binaries. His main objective, however, was to determine, as he put it, 'the construction of the heavens'. One big problem faced him—that of finding the distances of the stars. He had no way of telling whether faint stars were farther away than bright ones. Not to be outdone, he assumed

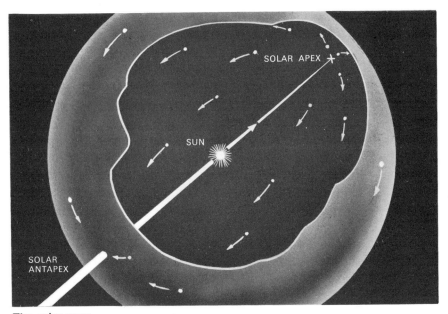

The solar apex.

that they were. All stars, he thought, were similar in actual brightness, or as we would now say, in *luminosity*. Brightness therefore became a measure of distance. He put *Sirius* at a unit distance of just over 38 million million miles—quite a good guess in view of the modern value of about 52 million million. But it so happens that stars have a considerable range in luminosity. *Rigel* and *Arcturus*, for instance are about equal in brightness, but *Rigel* is over 23 times more distant than *Arcturus*. *Rigel* must therefore be by far the more luminous of the two. Even so, faint stars are *on the average* farther away than bright ones, so Herschel's assumption was not a bad one after all.

Herschel also assumed that the stars were uniformly scattered throughout space. This again, as he found later, is not so. But by adopting it and counting the number of stars visible in his telescope when it was turned to different parts of the sky, he concluded that the Milky Way was not just an uneven band or belt of stars but an optical effect. It owed its brightness to vast numbers of very faint (and hence very distant) stars. Moreover, these stars were scattered to greater depths in space than those in other directions. This led directly to the idea that the sun lay near the centre of a flattened bun-shaped system of millions of stars. The system reached out farthest in various directions along the Milky Way and had a maximum radius of not less than 497 times the distance of *Sirius* from the earth. The Milky Way therefore defined the general plane of a vast system of stars known later as the *Galaxy*.

The distance of a star other than the sun first became known in 1838. In that year the German astronomer F. G. Wilhelm Struve announced that he had found a parallax of $0''\cdot 261$ of arc for the bright star *Vega*. The star lies beside a faint one which is really much farther away than *Vega* itself. Struve therefore succeeded where Herschel had failed, but since his measurements were rather few and far between the result didn't inspire much confidence. Later in the same year another German astronomer, Freidrich W. Bessel, announced that the star 61 *Cygni* had a parallax of $0''\cdot 3136$ of arc. Although by no means a bright star its proper motion indicated that it was fairly near. Bessel kept it under regular observation for a year, using for the purpose a heliometer made by Fraunhofer. In 1840, after making another series of measurements, he increased the parallax to $0''\cdot 3483$ of arc. Meanwhile, a third astronomer, Thomas Henderson, announced that his observations of the star *Alpha Centauri*, made at the Royal Observatory, Cape of Good Hope, showed that it had a parallax of $1''\cdot 0$ of arc.

Modern estimates of the annual parallaxes of *Vega*, 61 *Cygni* and *Alpha Centauri* are $0''\cdot 123$, $0''\cdot 292$ and $0''\cdot 751$ of arc respectively. They are the result of measurements made not at the telescope but of photographs taken with the telescope. Just what any one parallax corresponds to in terms of distance depends, of course, on the length of the base line, that is, on the value of the astronomical unit. On the basis that this is roughly 93 million miles, the distances of even the nearest stars come to many million million miles—to numbers so big that it is almost impossible to imagine them. Astronomers therefore use two units much larger than the mile—the *light-year* and the *parsec*. The light-year is the distance light travels in a year. Since its velocity is 186,282 miles a second, light travels about 6 million million miles in a year. Using this unit the distances of *Vega*, 61 *Cygni* and *Alpha Centauri* come to 26, 11·2 and 4·3 light-years respectively. The parsec, the distance of a star which has a parallax of 1"·0 of arc, is equal to 206,265 astronomical units or 3·259 light-years.

Henderson chose wisely when he selected *Alpha Centauri* since it is almost the nearest star. Actually it is a system of three stars. Two of them form *Alpha Centauri*, itself a binary system. The third, *Proxima Centauri*, a faint star fairly near *Alpha*, is the nearest star. Its parallax of 0"·762 of arc corresponds to a distance of just under 4·3 light-years.

Although the smallness of the angles of parallax came as no surprise they were, in a sense, disappointing. They showed that the trigonometrical method would be quite useless for finding the distances of the stars in the Milky Way, and hence, the size of the Galaxy. Even today, with the help of photography and all manner of refinements to observation, the trigonometrical method is reliable only out to about 300 light-years. Many bright stars lie well beyond this range. But once a star's distance is known its luminosity follows straightway. One of the laws of light is that the brightness of a point source (e.g., a star) falls off as the square of its distance from the observer. This means that if the distance of, say, *Sirius* were doubled, its brightness would be only one quarter of what it is now. For treble the distance the brightness would be one ninth ... and so on. Calculation shows that *Sirius*, with an apparent magnitude of −1·47 and at a distance of 8·7 light-years, has a luminosity 26 times that of the sun.

The term *apparent magnitude* refers to a way of grading the stars according to their brightness. It was introduced by Hipparchus who divided the stars into six classes or magnitudes. The brightest were said to be of the first magnitude, the next brightest of the second magnitude, and so on to those of the sixth magnitude, or stars just visible to the unaided eye. A star of the first magnitude is now reckoned to be 100 times brighter than one of the sixth. A difference of **one magnitude** corresponds to a brightness ratio of 2·512. That is, a **star of magnitude** 1·0 is 2·512 times brighter than one of magnitude 2·0.

Comparison of the brightnesses of different stars can be done with instruments called *visual photometers*. They show that several stars are brighter than magnitude 1·0. *Altair*, for instance is 0·75, *Capella* 0·09 and *Sirius* —1·47, or 1·47 magnitudes below zero magnitude. These numbers represent *visual magnitudes*. Another method is to measure the sizes of star images on photographic plates. These can then be compared with those of stars of known brightness and expressed as *photographic magnitudes*. A third and extremely accurate way is to obtain *photoelectric magnitudes* by means of a small but very sensitive photoelectric cell. Starlight falling on the cell releases electrons which then form an exceedingly weak electric current. The number of electrons released and hence the strength of the current is directly proportional to the intensity of the light falling on the cell. In practice the current is amplified and recorded by the deflection of a pen moving across a roll of paper. The magnitude of a star can therefore have different values for one and the same star depending on how it has been measured. The general custom, however, is to use the term on its own whenever visual magnitude is being referred to.

Since stars lie at different distances, their luminosities can best be compared if we imagine them to be all at one particular distance. The one usually adopted is 10 parsecs or 32·6 light-years. It corresponds to a parallax of $0''\cdot 1$ of arc, and magnitudes calculated for this distance are called *absolute magnitudes*. When this is done it is found that the range in luminosity is immense. Among the 20 brightest stars it varies from 50,000 for *Deneb* to 1 for the sun. Yet some stars are much more luminous even than *Deneb* and the sun is over a million times more luminous than certain nearby small stars. As will be seen later, this great range is due to two features—surface area and temperature. The sun, with an absolute magnitude of 4·86, is about average in size, temperature and luminosity.

To determine the distances of some very remote objects astronomers can rely on a class of variable star known as *cepheids*. The term 'variable' means that they rise and fall in brightness and therefore in luminosity. They are called 'cepheids' after *Delta Cephei*, a typical member of their class. In nearly $5\frac{1}{2}$ days *Delta Cephei* brightens fairly rapidly from magnitude 4·5 to 3·7 and then gradually fades back again. The time of one complete light change is called its *period* and this, together with the rate and extent of the change, remains practically constant. The period

114 · BOOK OF ASTRONOMY

does, however, vary from one cepheid to another, although the general pattern of light change of *Delta Cephei* is characteristic of them all.

In 1912 Henrietta S. Leavitt discovered that there was a definite relationship between the period of a cepheid and its luminosity. Her study of photographs of the Small Magellanic Cloud (a cloud of stars in southern skies now known to be another galaxy) showed that it contained many cepheids. She also found that when their periods were plotted one by one against their (average) magnitudes, the points obtained lay on a simple curve. Bright cepheids had periods of several days, faint ones had periods of a day or so. Now the Small Magellanic Cloud is so far away that for all practical purposes its cepheids are at the same distance from us. The period of any one of them could therefore be found merely by measuring its average magnitude. The next thing was to determine the distances and hence absolute magnitudes of a number of near cepheids. When this was done (from studies of proper motions) a curve could be constructed to show the relationship between period and absolute magnitude

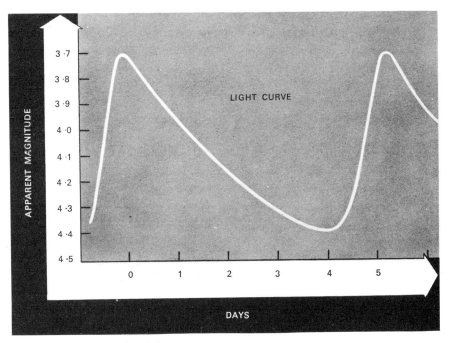

Light curve of Delta Cephei.

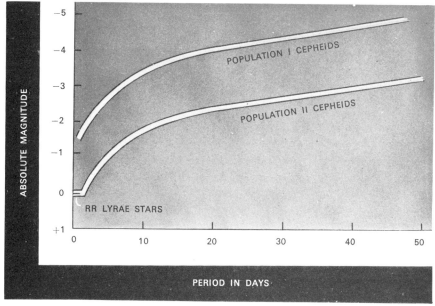

Cepheid period-luminosity relationship.

(or luminosity) for all cepheids. Always assuming, of course, that a cepheid in *any* part of the sky kept to the period-luminosity relationship. So once the period of a cepheid was known, its absolute magnitude could be read off from the curve. It was then only necessary to determine the star's apparent magnitude in order to obtain its distance.

One of the main problems with cepheids is that even their nearest members are so far away that estimates of their distances tend to be unreliable. As a result, and until recently, astronomers had great difficulty in fixing the zero of period-luminosity curve. It is now known that some cepheids (Type I) are about 1·5 magnitudes brighter than others (Type II) of the same period. But all of them have at least one thing in common. The lines in their spectra show a slight Doppler-Fizeau effect. They swing slightly about a mean position in such a way as to indicate that cepheids are pulsating stars. Their spectra also indicate that they change in surface temperature with change in luminosity. It therefore appears that they brighten as they expand and get hotter, and fade as they contract and get cooler.

Another useful type of pulsating star for finding great distances are *RR Lyrae* stars, so named after the seventh magnitude star *RR Lyrae*, the brightest member. The pattern of their light changes is similar to that of the cepheids but the periods range from about an hour to a little over a day. Their value lies in the fact that they all have an absolute magnitude of 0·5 and are therefore about equal in luminosity. So once an *RR Lyrae* star gives itself away by the nature of its light change, its distance can be found from its apparent magnitude.

Cepheids were first put to work as distance indicators by the American astronomer Harlow Shapley. In 1916, using the 60-inch reflector of the Mount Wilson Observatory, he began to study the distribution in space of objects known as *globular star clusters*. These are compact swarms of thousands of stars so remote that in small telescopes they look like faint tail-less comets. Each swarm forms a tightly-knit and almost spherical ball of highly luminous stars. Shapley found that many of these clusters contained either cepheids or *RR Lyrae* stars, and sometimes both. Others, so distant that they could not be resolved into stars, had to be judged by their apparent sizes and brightness. Shapley assumed that one which looked very small and faint was more distant than one which looked larger and brighter. As a result he obtained the first evidence of the vast extent of the Galaxy. Even the nearest cluster, an object known as *Omega Centauri*, turned out to be 22,000 light-years away, while the most distant was 230,000 light-years.

Shapley also found that while the 100 known globular star clusters appeared to be more or less equally distributed on either side of the plane of the Milky Way, one-third of them occupied a region in Sagittarius. Their actual distribution in space, however, was such that they formed an immense roughly spherical system whose centre lay in the direction of Sagittarius. This centre, he suggested, coincided with that of the Galaxy. So the sun did not occupy a central position in the Galaxy after all. Instead, it lay some 60,000 light-years from a centre in the direction of Sagittarius. As for the Galaxy, this had, in his opinion, the flattened shape of a lens, 300,000 light-years in diameter and 10,000 light-years in thickness. Small wonder that many astronomers found his figures difficult to grasp and viewed the entire picture with great suspicion!

The flattened shape of the Galaxy indicated that it was a rotating system. Further support for this idea was the fact that the sun and other nearby stars were moving through space. But if Shapley's figures could be relied on, the sun, with its speed of about 12 miles a second towards the solar apex, would require at least 4,500 million years to revolve once around the Galactic centre. Of course, if the Galaxy rotated as one whole piece in the manner of a cartwheel, all the stars near the sun would go along together like specks of dust on a moving wheel-spoke. In addition, stars near the rim would travel much faster than those near the centre.

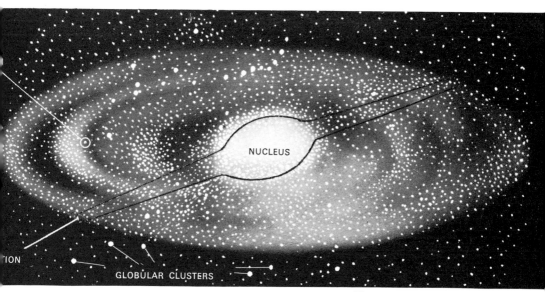

Shape of the Galaxy.

In 1927 the Dutch astronomer Jan H. Oort announced that this view was incorrect. His studies of the radial velocities and proper motions of a large number of highly luminous stars revealed that those nearer to the centre than the sun forged ahead of the sun, while those farther away lagged behind. The stars in the Galaxy, he explained, moved around the Galactic centre in much the same way as the planets moved around the sun. Those near the centre travelled faster than those farther away. According to his calculations the sun had an orbital speed of about 140 miles a second and was about 27,000 light-years from the centre. All this meant, of course, that a great part of the total mass of the Galaxy, probably in the form of a fairly dense swarm of stars, was concentrated at and near its centre.

Meanwhile, Shapley had reduced the diameter of the Galaxy from 300,000 to 180,000 light-years. Robert J. Trumpler of the Lick Observatory had been at work in the 1920s studying the sizes and distance of several hundred fairly loose groupings of stars known as *open star clusters*. To determine their distances he applied the usual tests and found, much to his surprise, that their diameters increased with distance. This highly improbable result led him to the important discovery that some of the light of the clusters was lost on the way. Space, it seemed, was not empty but contained a haze which dimmed and reddened the light of very distant stars and made them appear fainter and cooler than they really were. Trumpler's observations showed that the haze lay in and near the plane of the Galaxy (the *Galactic plane*) and formed a layer similar to that of thick jam in a sponge-cake. The fact that it reddened light passing through it meant that it was made up of dust-like particles, widely spaced but spread over immense distances. The particles scattered the blue and violet parts of the light in much the same way as dust and molecules of air scatter sunlight and make the setting sun look red and the sky blue.

In the general haze were condensations in the form of great clouds of interstellar dust. These either weakened the light of the stars behind them, or, more often than not, cut it off altogether. Dust clouds fairly near the sun appeared on photographs of the Milky Way as empty spaces called *dark nebulae*. Those more distant had sprinklings of foreground stars, while the very distant ones were so well covered over by stars as almost to escape notice. In the direction of Sagittarius they formed a barrier so dense as to cut off most of the light from the central regions of the Galaxy. When Shapley took the effect of the haze and dust clouds into account the Galaxy shrank considerably. It was also found that the haze dimmed and reddened the light from globular star clusters (and therefore their cepheids and *RR Lyrae* stars) so that they too appeared farther away than they really were. The effect was most prominent for clusters whose direction lay near the plane of the Galaxy, that is, near the Milky Way. When this was allowed for, the diameter of the Galaxy shrank still farther to about 90,000 light-years. And there, much to everyone's relief, it stayed. For the sun's distance from the centre came to about 30,000 light-years, in good agreement with Oort's result.

The Galaxy is now thought to contain nearly 100,000 million stars. Of these several thousand million can be photographed with big telescopes but only about 2,500 can be seen with the unaided eye on a clear night. In its general shape the Galaxy is rather like an enormous fried egg—round, and fairly flat except for a bulge in the middle. The bulge, called the *nucleus*, contains millions of high-luminosity stars fairly densely packed but free from interstellar dust. It cannot be seen with optical telescopes, but radio telescopes can penetrate the dust clouds and reach its outer parts.

The overall diameter of the Galaxy is about 100 thousand light-years, but since, like the earth's atmosphere, it has no definite limit, this is only a rough estimate. Its average thickness is about 5,000 light-years. The sun, roughly 33,000 light-years from the centre, has an orbital speed of approximately 160 miles a second and makes one revolution in about 220 million years. These figures will undoubtedly be changed slightly as further observations on a wide front bring in new knowledge. But whatever the change one thing is definite enough—the sun, one star in a system of many thousand million stars, is a long way from the centre of the system.

The immense size of the Galaxy can perhaps best be pictured by thinking in terms of a model. Suppose we reduce the entire solar system

Horsehead Nebula, a dark obscuring mass of interstellar dust in Orion.

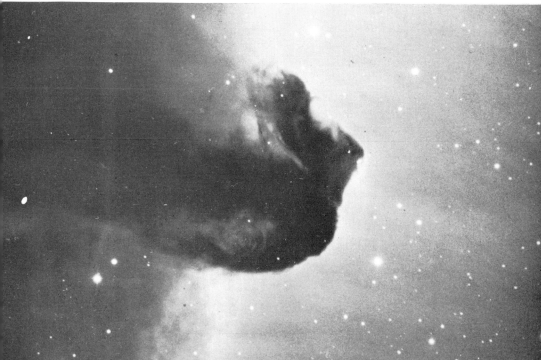

to the size of a pinhead. Then the nearest star, *Proxima Centauri*, is about a quarter of a mile away. All the naked-eye stars lie within a distance of about 70 miles, but the distance right across the Galaxy almost equals the diameter of the earth. Indeed, on this scale the Galaxy, together with its surrounding system of at least 100 globular star clusters, just about fills the earth.

One of the most remarkable discoveries in astronomy in recent years is that the Galaxy has an overall spiral structure. Its stars, instead of being uniformly scattered like currants in a well-made bun, are concentrated in long curved branches or arms. Seen from a great distance it would have a definite whorl-like appearance similar to that of a Catherine-wheel. Parts of a few arms can be traced by plotting the positions of highly-luminous blue-white stars. But the best results are given by studies of the distribution of hydrogen gas.

Radio telescopes show that, in addition to dust-like particles, interstellar space also contains thinly-spread neutral hydrogen gas. The term 'neutral' means that every hydrogen atom is complete, that is, made up of a single proton and single electron. Its energy state is low but once every few million years, on average, the electron changes its direction of spin. The atom then either absorbs or emits an exceedingly small amount of energy. A gas made up of neutral hydrogen emits at a wavelength of 21 cm. It also reveals its motion relative to the observer on earth by a Doppler-Fizeau shift of this line. So by measuring the amount of shift, astronomers can obtain the velocity of approach or recession of the gas itself. The radial velocities are then converted into distances by comparing them with those one might expect to find in corresponding parts of the Galaxy. Studies of the amount of gas, indicated by the strength of the emission, show that it is concentrated in the spiral arms. A map of the Galaxy can therefore be drawn which shows at once the distribution of neutral hydrogen and the main spiral features. The great value of the radio method is that it can extend our surveys far beyond the optical range. For while light is usually completely scattered and absorbed by clouds of interstellar dust, 21 cm. radio waves pass through them without appreciable loss.

CHAPTER 12

Gas, Dust and Stars

BETWEEN 1771 and 1784 the French astronomer Charles Messier, searching the sky for comets, came across a number of misty patches of light which remained fixed among the stars. Some of them proved to be clusters of faint stars, others showed no traces of stars. Altogether he catalogued 103 of these objects which he divided into two groups—star clusters and nebulae. As a result, the brighter members of his list are usually known by their *Messier numbers*. For instance, Messier 13, or M13 for short, is the globular star cluster in Hercules, an object discovered by Halley in 1714 and the brightest of its kind in the northern sky: M31 is the Great Nebula in Andromeda, noticed in 1612 by Simon Marius, a contemporary of Galileo: M42 is the Great Nebula in Orion, a diffuse patch of light centred on the star *Theta Orionis* and discovered in 1656 by Huygens.

During his explorations of the sky Herschel came across and eventually catalogued no less than 2,500 star clusters and nebulae. His telescopes resolved so many nebulae into stars that at first he thought they were all starry in nature. But as time went on others just wouldn't be resolved, not even with the 48-inch telescope, so he concluded that they were composed of a 'shining fluid' or 'milky nebulosity'. Later astronomers did not agree on this point. Lord Rosse, for one, claimed that no real nebula existed—every one seemed to be a cluster of stars. This was no idle guess but a direct result of observations made in 1845 with a giant 60-foot reflector of 72 inches aperture. Even the Great Nebula in Orion, regarded by Herschel as a 'shining fluid', appeared to be on the verge of resolution into stars. And there the matter rested until 1864 when Huggins announced that eight nebulae gave bright-line spectra. They could not therefore be very remote collections of stars but masses of shining gases. The spectroscope also showed that the Great Nebula in Orion had a gaseous nature. On the other hand, the Great Nebula in Andromeda gave a continuous spectrum, thereby indicating that it was probably made up entirely of stars.

The eight objects to which Huggins directed his spectroscope were all *planetary nebulae*. They were first classified as such by Herschel who commented on their faint, greenish colour and small, roundish shapes. About 500 are now known. Each one, it seems, consists of a roughly spherical mass of extremely thin gases, with one or more small but intensely hot stars at its centre. The gas mainly responsible for the light is doubly ionised oxygen, or oxygen with all of its atoms stripped of two electrons. This gives the nebula its greenish colour. Other gases are neutral and ionised hydrogen, and ionised neon, helium and nitrogen. They all shine by absorbing the intense ultra-violet radiation of the central star and sending it out at various longer wavelengths.

This process, called *fluorescence*, is also found to be going on in most other nebulae. On their own the gases would not shine at all: they owe their light solely to the hot stars embedded in them.

Some planetary nebulae, like M67 in Lyra, have a definite ring-like appearance. This is no more than an effect of perspective. In reality the surrounding gases form a shell (sometimes two and three shells) round the central star. This strongly suggests that the shell has been blown off

Planetary nebula in Aquarius.

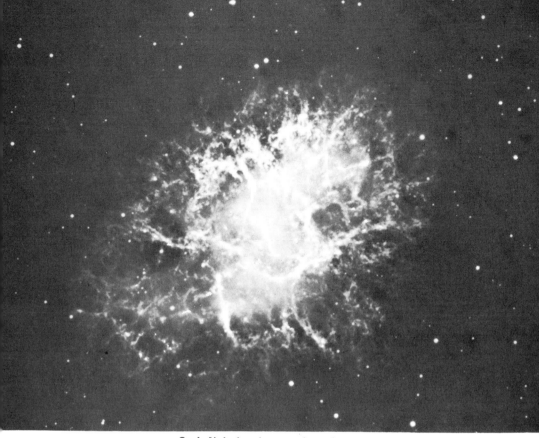

Crab Nebula, the remains of the supernova of 1054.

from the central star. Astronomers have therefore compared the sizes of certain planetary nebulae on modern photographs with those on photographs taken some sixty years or so ago. In some cases it is clear that the nebulae have got perceptibly larger. This is confirmed by spectrograms: measurements of the Doppler-Fizeau shift of the spectrum lines indicate that the shells are moving at many miles a second. Finally, planetary nebulae look small only because they are so far away. On the average they have diameters of about 0·5 light-year but large ones may extend to more than a light-year.

The most remarkable planetary nebula is M1, the Crab Nebula. So remarkable, in fact, that it practically stands in a class of its own. It lies in the constellation of Taurus in the place where Chinese astronomers in A.D. 1054 saw a new star appear. According to their records the star rose to equal Venus in brightness and then slowly faded to disappear completely. There is now little doubt that M1 represents the remains of a supernova. Spectrograms and photographs reveal that it is a mass of gas in a state of tremendous upheaval. They also show that the gas is expanding at a rate of some 600 miles a second. Knowing this rate, and assuming that it has always been fairly constant, it is possible to find when all the material was concentrated in one spot. The calculation shows that the star exploded some 900 years ago, that is, around 1054.

Whether the core of the original star still remains intact is uncertain. The star's outer parts, or perhaps all of it, is (or was) flung over a volume of space some 5 light-years across. We add 'or was' since M1 is believed to be about 3,500 light-years away. This means that the explosion did not occur in 1054 but around 2500 B.C. Also, what we now find out about M1 is already some 3,500 years out of date. To appear as bright as Venus, a star at this great distance must have risen in luminosity to equal about 100 million suns. If this ever happened to the sun, the earth and every other solid object in the solar system would swiftly be turned into gas. Fortunately for us the sun is too steady a star to destroy itself in this way. As far as we can tell it will continue to maintain its present output of light and heat without much change for at least another 2,000 million years.

M1 is also an extremely powerful source of radio waves. The radio energy is thought to be produced by extremely fast-moving and therefore highly-energetic electrons which spiral around lines of magnetic force. Indeed, recent optical studies of a special nature suggest that the greater part of the radiation, both radio and optical, is due to this process. Incidentally, the remains of the supernovae of 1572 and 1604 are very poor affairs compared with M1. The former has no visible remnant but the latter shows a few nebulous fragments. Both sites emit radio waves.

While planetary nebulae look fairly small and compact, others, like M42, the Great Nebula in Orion, are large objects with no definite boundaries. These *irregular nebulae*, as they are called, shine by fluorescence and owe most of their light to ionised oxygen and hydrogen. The bright lines of their spectra sometimes appear on faint continuous backgrounds. This indicates that they also reflect and scatter some of the light of the stars nearby. The reflection comes from clouds of interstellar dust associated with the gas. In many cases parts of these dust clouds appear as dark rifts and gaps in the shining gas, while the spectrum shows features due to both the emission of light (bright lines) and its reflection (con-

Great Nebula in Orion.

tinuous background). This is certainly true of M42, the brightest part of a vast complex of gas and dust which extends over the entire constellation of Orion. A medium-sized telescope shows that it owes most of its light to a group of four stars known as the *Trapezium*. On photographs the group is generally lost in the bright light of the nebula itself. But since long-exposure photographs can record the extremely faint glow of the nebula's outer parts, they enable us to appreciate its immense size and complexity. Estimates of its distance vary between 900 and 1,800 light-years, but recent work suggests that it is somewhere near the upper limit.

Some nebulae called *reflection nebulae*, shine solely by reflected and scattered light. The outstanding example of this type is found in the Pleiades, a well-known cluster of stars about 400 light-years away. Long-exposure photographs reveal that its five brightest stars are wreathed in wisps of nebulosity. But since the spectrum of the nebulosity is similar to that of the stars involved in it, the reflection and scattering of starlight by dust particles is clearly indicated.

Recent studies of the distribution of planetary nebulae show that most of them lie in the direction of the centre of the Galaxy. They also tend to crowd in the region of the Galactic plane, although several lie well above and below it. It therefore seems that they do not belong to the spiral arms. If some happen to be in the arms it does not follow that, like the sun and its neighbours, they were 'born and bred' there. Irregular nebulae, on the other hand, keep to the Milky Way and are scattered fairly evenly along its course. In this respect they are like the dark nebulae, which, as we have seen, are clouds in a general haze of interstellar dust lying in and near the Galactic plane. They appear dark when they cut off the light of the stars behind them, and bright when they reflect and scatter the light of stars nearby. In a similar way, irregular nebulae are shining clouds in a layer of interstellar gas which also lies in the Galactic plane.

While the general haze of dust dims and reddens the light of distant stars, the gas absorbs only particular wavelengths. Its presence was first detected in 1904 by the German astronomer G. Hartmann. Spectrograms of the star *Delta Orionis* showed a certain line due to calcium which could not be produced by absorption in the star's atmosphere. The line was thin and dark whereas the others were broad and fuzzy. Later observations showed that the calcium was interstellar, that is, it occupied the vast space between *Delta Orionis* and the solar system. They also showed that other

The Double Star Cluster in Perseus.

gases, notably hydrogen and helium, with small amounts of calcium, oxygen, sodium and other elements, were also present in space.

Also members of the Galaxy, and showing a tendency to lie in its plane, are the 'open' star clusters or *Galactic clusters*. The Pleiades, the Hyades and Coma Berenices are well-known naked-eye examples. Although several of their brighter stars can be seen without optical aid they appear at their best through binoculars or a low-power telescope. Other clusters, like Praesepe in the Crab and the double cluster in Perseus, appear as misty patches to the naked eye but show their stars in small telescopes. Others again, even more distant, reveal their starry nature only in fairly large telescopes. Over 300 are known, but there are probably many more hidden behind the dark dust clouds of the Milky Way.

Proper motion and radial velocity observations show that a Galactic cluster is a moving family of suns. Its stars all travel at the same speed in the same direction, moving as a unit around the centre of the Galaxy and also with respect to the sun. For a near cluster like the Hyades, 130 light-years away, the proper motions are fairly large. They can be used to build up a picture of the cluster, discover the kinds of stars it contains and how they are arranged in space. They also enable astronomers to pick out the actual members of the cluster from stars which happen to be in the line of sight. Although the bright star *Aldebaran*, for example, looks as though it belongs to the Hyades, it is really a foreground star at a distance of about 65 light-years. The Hyades and Pleiades each contain at least 200 stars. Each cluster occupies a volume of space about 45 light-years across and has its stars slightly more crowded at the centre than at the edge. It is therefore a very empty affair, with one star separated from another, on average, by several light-years.

Through proper motion studies, five of the seven stars of the Plough have been found to be members of a group of at least 40 stars. *Sirius*, about 90° away from the Plough, also belongs to the group. In other words, the cluster is so 'open' and so near, that, apart from the proper motions of its members, it fails to appear as a cluster. If all seven stars of the Plough belonged to the group they would keep their present relative positions and distances for thousands of years. As it is, five move together in one direction while the other two move in different directions. So in the far distant past the stars did not form a plough-shaped figure, nor will they do so in the far distant future.

While a Galactic cluster has only a few hundred stars, a globular star cluster contains thousands and sometimes hundreds of thousands. It is ball-like in shape, usually slightly flattened, with its stars greatly crowded towards the centre. Only low-exposure photographs can show the individual stars at and near the centre. As soon as the exposure time is increased to record more outlying stars, the central part gets over-exposed and appears on the photograph as a more-or-less uniform blob. This partly explains why it is difficult to estimate the number of stars in a rich globular cluster. One has also to remember that the cluster is, so to speak, seen edgewise-on. The apparent closeness of its stars does not therefore give a proper indication of their actual arrangement. The nearest cluster, *Omega Centauri*, is about 22,000 light-years away and has a diameter of several hundred light years. It probably contains about a million stars, yet to the naked eye appears no brighter than a star of the fourth magnitude. Another fine object is 47 *Toucani*, also in the southern hemisphere. Northern observers can be proud of M13, the globular cluster in Hercules. It is about 34,000 light-years away, has a diameter of some 160 light-years, and probably contains well over 100,000 stars. What a sight it would be were its distance reduced to a mere 100 light-years! Its stars would then more than fill the sky and turn night almost into day. If such had been the case one wonders just what sort of a universe astronomers would have produced. At any rate, they would, in being placed so near the cluster, be well placed for getting a bird's-eye view of the entire Galaxy. Unless, of course, the cluster happened to be in the way!

As Shapley discovered, the system of about 100 known clusters appears to form a kind of framework around the Galaxy. Yet the framework is by no means fixed. Every star and cluster of stars in the Galaxy moves around the Galactic centre. This holds good for the mighty globular clusters as well as for the modest open clusters and individual stars like the sun. But while the orbits of open clusters, stars and other objects such as bright and dark nebulae, are almost circular and lie in and near the

M13, globular star cluster in Hercules.

M51, a spiral galaxy seen face-on.   **Spiral galaxy in Coma Berenices seen edge-on.**

Galactic plane, those of the globular clusters are highly eccentric ellipses inclined at all angles to the Galactic plane. Moreover, radial velocity and proper motion measurements show that globular star clusters do not share in the general rotation of the Galaxy but proceed on their own individual ways with fairly low orbital velocities. Add to this the fact that their brightest stars are reddish and high in luminosity and it becomes evident that they form a distinct class. Distinct, that is, from objects in the spiral arms in organisation, motion and even age.

Until about fifty years ago astronomers could not agree about the precise nature of several objects in Messier's catalogue. Very little was known about even the brightest of them, M31, the Great Nebula in Andromeda. True, it gave a continuous spectrum crossed by dark lines, but then so did reflection nebulae. In 1917, however, the American astronomer George W. Ritchey discovered that novae had been recorded from time to time on photographs of M31. Their magnitudes, compared with those of novae at known distances in the Galaxy, indicated that the nebula was about a million light-years away. A few years later, in 1923-4, Edwin P. Hubble, using the 100-inch reflector of the Mount Wilson

Observatory, succeeded in resolving the outer parts of M31 and M33 (a nebula in Triangulum) into stars. To his great joy he found cepheids among them, and by using the period-luminosity relationship, was able to give both objects a distance of about 870,000 light-years. M31, he thought, had an apparent diameter of nearly 3°, so at its great distance the actual diameter was about 50,000 light-years. M33 turned out to be smaller,

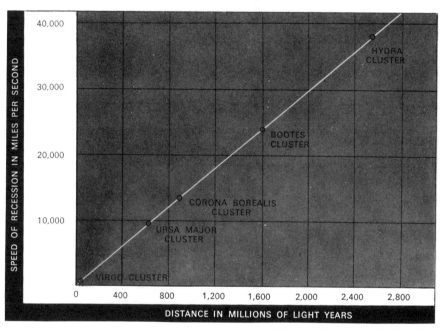

Velocity-distance relationship.

but still very large compared with star distances in the Galaxy. The conclusion was an obvious one. M31 and M33 were galaxies, independent of and external to our own Galaxy. The Galaxy was not the universe after all, but apparently just one system of stars, gas and dust in a universe of similar systems.

Just as Herschel pioneered in the realm of the stars, so Hubble pioneered in the universe of galaxies. It had been known for some time that there were thousands of small, faint, regular-shaped nebulae. The remarkable thing about them was that they occurred in all parts of the sky except the Milky Way. Why they did so became clear when Trumpler discovered the effects of interstellar absorption. Dust clouds in the Galactic plane completely cut off their light. Hubble found that they were all galaxies. The 100-inch telescope resolved several into stars and showed that the majority, M31 and M33 included, had a spiral structure. The spirals appeared at all angles. Some, like M51 in *Canes Venatici*, were practically face-on. Others, sideways-on, looked like spindles, each with a nucleus or bulge in the middle. In the case of galaxies too distant to show any stars, Hubble assumed that they were equally luminous. He then used their apparent magnitudes as rough guides to their distances and, by so doing, reached out to galaxies 130 million light-years away. He also found that galaxies, like stars, were not uniformly scattered in space. They had a tendency to occur in groups or clusters. The largest, and the nearest, in *Virgo*, contained several hundred members while one much more distant, in *Ursa Major*, contained about 300.

In 1929, Hubble came across a completely unexpected relationship. He had in front of him the estimated distances of 19 galaxies. He also had their radial velocities and those of 27 others. The velocities, of course, were obtained by measuring the shift in the spectrum lines due to the Doppler-Fizeau effect. Each one, when corrected for the sun's orbital motion about the centre of the Galaxy, turned out to be a velocity of recession. The shift was towards the red, hence the popular but misleading expression: 'the red-shift of the galaxies'. When he compared the two sets of figures he found, much to his surprise, that the velocities were roughly proportional to the distances. In other words, when the velocities corresponding to certain distances were plotted on a graph the points obtained fell on or near a straight line. A relationship of this kind is called a straight-line or linear relationship. Since it concerns the change in velocity with distance it is also known as the velocity-distance relationship. In popular language it means that the galaxies are all travelling away from us and from one another at speeds which increase as their distances increase.

CHAPTER 13

Stars and Galaxies

About a century ago, astronomers at the Harvard College Observatory made a spectrographic survey of all the brighter stars in the sky. As a result they found that stars could be arranged in a definite order or sequence of spectral types. Each type is represented by a letter of the alphabet and the main ones form the order O, B, A, F, G, K, M, R, N, S. The sequence also represents an order of decreasing temperature and therefore of change in colour. Stars of spectral types O, B and A are extremely hot and blue-white in colour, those of types F, G and K have moderate temperatures and are yellowish, while those of types M, R, N and S are cool, red stars. The temperature of a star's atmosphere (usually called the surface temperature) can therefore be judged from its spectrum and also from its colour.

The scale of surface temperature among the stars is quite wide. Near the top end are blue-white stars like *Rigel* and *Spica*. Their temperatures are about 25,000° C. and their spectra show hydrogen and helium absorption lines characteristic of B-type stars. Near the bottom end are red stars like *Betelgeuse* and *Antares*. Their temperatures, about 3,500° C., are low enough for the atoms in their atmospheres to combine and form molecules. Their spectra therefore show absorption bands due to titanium oxide and compounds of carbon. The sun, a yellowish star, is much nearer the bottom end of the scale than the top. Its surface temperature of 6,000°C. and the great number of lines in its spectrum due to metals like iron, calcium, magnesium and titanium, brand it as a G-type star.

As far as its colour is concerned a star is like a hot poker. The more the poker is heated the higher its temperature gets, and in the process, its colour changes from dull to bright red, then to orange, yellow, and pale yellow through to white and blue-white. White-hot pokers are much brighter (that is, more luminous) than red ones, and the same is true of the stars. Further, a large poker will be brighter than a small one at the same temperature because of its greater surface area. This also is true of stars.

In their case, however, it is interesting to compare luminosity (or absolute magnitude) with temperature (or spectral type).

This was first done by Enjar Hertzsprung of Denmark in 1905, and in greater detail, by Henry Norris Russell of Princeton University, U.S.A., in 1913. Russell had reliable parallaxes (and therefore reliable absolute magnitudes) for only about 300 stars. Even so, the comparison showed that most of the stars, the sun included, formed a definite sequence, known as the *main sequence*, or *dwarf sequence*. All along the sequence, luminosity increased with temperature, which, after all, is what one would expect it to do. The surprising thing was that the other stars kept well outside this trend by having very high luminosities for their spectral types. *Betelgeuse*, for example, a cool star of spectral type M, had a luminosity far and away above that of a star of the same spectral type in the main sequence. It was as if there were two pokers, both red, but one immensely brighter than the other. There could be only one possible explanation. *Betelgeuse* was a giant compared with M-type stars in the main sequence. A similar although slightly smaller difference in luminosity was also found in orange K-type and yellow G-type stars. They too had their giants and their dwarfs.

In 1914, Walter S. Adams of the Mount Wilson Observatory, followed up Russell's work with an important discovery. He found that it was possible to tell whether a star was a giant or a dwarf merely by measuring the relative intensities of certain pairs of lines in its spectrum. The measurements could also be used to determine its luminosity and hence, from its apparent magnitude, the distance or *spectroscopic parallax.* As a result the luminosities and spectral types of many more stars could be compared. Spectroscopic parallaxes also helped astronomers to trace out the main features of the Galaxy to distances of several thousand light-years. Another great advance came in the 1920s when astronomers at the Mount Wilson Observatory successfully measured the angular diameters of a few giant stars. They attached a special arrangement of mirrors to the 100-inch telescope, so turning it into what is known as an *optical interferometer*. When the angles, all extremely small, were changed into distances, it was clear that if some stars were giants, *Antares* and *Betelgeuse* were supergiants. *Betelgeuse*, already known to vary slightly in brightness, was also found to have a variable diameter.

Today, after an immense amount of further work, astronomers find that most stars still fall into one of two broad groups—main sequence and giants. In the giant group are stars like *Arcturus* and *Aldebaran*, much larger than the sun and about 100 times more luminous. The group is now extended into the *supergiants* by highly luminous stars like *Betelgeuse* and *Antares*. *Betelgeuse* is one of the largest known stars. Its diameter varies from 300 to 420 times that of the sun and it is some 13,000 times more luminous. At its largest it could easily contain the entire orbit of Mars. *Antares*, slightly smaller, has a diameter about 285 times that of the sun. Yet because of their great distances, both stars appear as points of light even in the 200-inch telescope. A number of stars of other spectral types also come in the supergiant group. They have very high luminosities but none can compare in size with their red classmates.

Another group is that of the *white dwarfs*. Compared with main sequence stars they are small in size and low in luminosity. The one best known is *Sirius B*, so named because *Sirius* is actually a binary system. *Sirius A*, about 26 times more luminous than the sun, is the bright star we see in the sky. *Sirius B* is faint, and owing to its very bright companion, difficult to see even in large telescopes. Its luminosity, less than $\frac{1}{100}$ that of the sun, is found from its absolute magnitude. Its surface temperature (revealed by its spectrum) then shows that it has $\frac{1}{50}$ the diameter of the sun. But what about its mass? This too can be obtained by working out the orbits of the two stars and applying Kepler's laws and Newton's inverse square law of gravitation. *Sirius B* then presents itself as a star about halfway between the earth and Uranus in size, yet equal to the sun in mass. Its average density turns out to be 500,000 times that of water. A cubic inch of its material, if brought to the surface of the earth, would probably weigh over 8 tons!

White dwarfs, all faint stars, are detected by their large proper motions. About 200 have been discovered, all of them near neighbours of the sun. A few are even smaller than the earth and one, LP768-500, of magnitude 18·2, is thought to be about half the size of the moon. If this is so, the average density of its material must be of the order of 100 million times that of water.

Mass, and therefore density, is an important property of stars. The only really satisfactory way of finding it is to study the orbits of visual binary stars. That is, the orbits of double stars, which like *Sirius*, appear as two separate stars in the telescope. The stars in many binaries are so close that together they look like a single star. But as they waltz around each other they give themselves away by a regular doubling of the lines in their combined spectrum. Double stars of this kind are called *spectroscopic binaries*. Among them are the bright naked-eye stars *Rigel, Capella, Spica,* and *Alpha Crucis* (a pair of spectroscopic binaries, that is, four stars in all). Some double stars are *eclipsing spectroscopic binaries*. The earth lies so near the planes of their orbits that one star regularly passes in front of the other. As a result the star (really two stars) goes through a rhythmic change in brightness. *Beta Persei* or *Algol* is one of these. It appears almost constant in brightness (magnitude 2·3) for about 2½ days. Then for 5 hours it fades to ¼ of this brightness and, after another 5 hours, returns to its former brilliance.

In only a few cases can the orbits of spectroscopic binaries be so fully determined as to give the individual masses of the stars. The others reveal only the mass of the system as a whole. In 1924, however, Sir Arthur Eddington discovered from theory a relation between the mass and

Giant and dwarf stars.

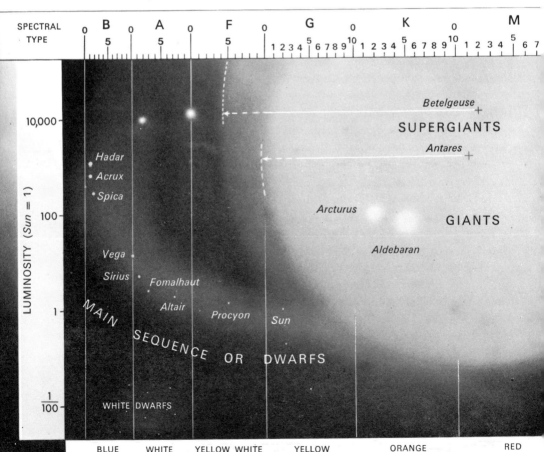

luminosity of a star. According to this, the greater the mass of a star the higher its luminosity. A star with twice the mass of the sun, for instance, has about 10 times its luminosity. The relation does not apply to the white dwarfs. So, knowing the mass and size of a star, it is a simple matter to work out its average density. The range in density is found to be very wide. At one extreme are the red supergiants, with their outer layers thinner than the material in a comet's tail. At the other are the white dwarfs. Their atoms, stripped of electrons, are so tightly packed that there is hardly any space at all between them.

Astronomers now have a pretty clear idea of the way stars shine. Discoveries in atomic physics have led to the idea of a star as a kind of enormous atomic power-station. Deep in its interior, under enormous pressures and at temperatures of several million degrees, hydrogen is first transmuted or changed into helium, which is then changed into carbon. The process, known as a *thermonuclear reaction*, produces vast quantities of energy which are poured into space as light and heat. Even the sun, a modest dwarf star, is thought to change 4 million tons of its mass into energy every second. But since the sun is so immense this loss is negligible.

The fact that every star is using up its store of hydrogen means that it cannot shine forever. Like every human being it must pass from birth, through youth and old age, to death; but in a period of time so vast that a human lifetime, in comparison, is less than a fleeting moment. Until recently, the problem facing astronomers was how to tell a young star from an old one. Did all stars begin as cool, red giants or supergiants, get hotter and smaller with advancing age, and then cool to reach the red dwarf stage of old age? If so, where did the white dwarfs, novae and variable stars fit into the picture?

An important clue to the problem came in 1943 when Walter Baade succeeded in photographing individual stars in the nucleus of M31, the Great Galaxy in Andromeda. Using the 100-inch reflector of the Mount Wilson Observatory, he found that photographic plates made sensitive to red light succeeded where all others had failed. Not only did they resolve the nucleus of M31 into stars but also two much smaller oval-shaped or *elliptical galaxies* near M31. He therefore concluded that the stars resolved in this way were highly luminous red giants. He also suggested that there are two populations of stars. 'Population I' is found in the arms of spiral galaxies and in less regular galaxies like the Magellanic Clouds.

It contains dwarf stars like the sun and has for its brightest members the very hot blue-white stars of open clusters and gaseous nebulae. 'Population II' is found in the central regions of spiral galaxies, in elliptical galaxies, and in globular star clusters. Its brightest members are cool but highly luminous red-giant stars.

According to modern ideas of stellar evolution, a star is formed out of a cloud of cold cosmic gas and dust. As the cloud contracts under its own gravity it gets hotter. It becomes a star when it is hot and compressed enough for hydrogen to start changing into helium. The star then settles down as a main sequence or dwarf object. Just how long it stays like this depends on its mass. A modest star like the sun can remain on the main sequence for more than 10,000 million years, but a massive, hot, and highly luminous star will stay there for only a few million. While on the main sequence the star remains much about the same in size and brightness. But when it has used up nearly all its hydrogen it swells and brightens to become a red giant. In the sun's case the swelling will go on for only a few million years. So, since the sun's age is reckoned to be about 8,000 million years, it should not change a great deal in the next 2,000 million.

What happens after the red-giant stage is only roughly known. Some stars become unstable and vary in their light (cepheids, *RR Lyrae* and other variable stars), some shed their surface layers (novae and planetary nebulae), others blow themselves up (supernovae). Stars similar in mass to the sun must eventually use up all their nuclear fuel, helium as well as hydrogen, and pass into a white-dwarf state. After that they rely on contraction alone for their light and heat, and finally, end as dark bodies.

If the life histories of stars are anything like this, the sun, along with other stars of Population I, must have evolved later in the history of the Galaxy than the stars of Population II. The very fact that Population I stars are found in regions rich in gas and dust supports this view. Population II stars, on the other hand, occur in regions free from dust. All the dust, presumably, has already been turned into stars. Population II stars would therefore appear to be a declining population. But whether Population I is increasing or decreasing we cannot, as yet, say. One thing is very clear. These modern ideas cut right across the old belief that the stars were all formed together and flung across the sky to remain there for ever. Instead, we now know that they were formed at different times as the Galaxy evolved. Also, that they will go on being formed in the spiral

M31, the Great Spiral Galaxy in Andromeda.

arms of the Galaxy long after many of those we now see have ceased to shine. In this respect stars are like people. Individuals come and go: some live to a good old age, others live for only a short time, but as long as birth can go on the population will continue.

In 1952 it became clear that all of Hubble's estimates for the distances of galaxies would have to be revised. Baade found that photographs of M31 taken with the 200-inch telescope on Palomar Mountain, failed to show any highly luminous *RR Lyrae* stars. If the galaxy were at a distance of 870,000 light-years (Hubble's first estimate), these stars ought to have appeared on the photographs. To make matters worse, Hubble had decided to reduce the distance to 750,000 light-years. Baade also found that the cepheid variables used by Hubble to obtain the distance of M31 were Population I stars, all 1·5 magnitudes more luminous than cepheids which, like *Delta Cephei*, belonged to Population II. As a result, Hubble had been misled into making the distance of M31 and other galaxies about 50% too small. On the old scale our Galaxy was larger than any other stellar system—a highly suspicious result. On the new scale, M31, twice as far away as before, doubled in size and became larger than the Galaxy.

M31 is now thought to be just over 2 million light-years away and to have a diameter of about 120,000 light-years. It looks very much as our Galaxy would look if it were seen from the outside, complete with its system of globular star clusters, but tilted about 15° from the edge-on position. Its spiral arms, outlined by dark clouds of dust, contain patches of glowing gas (irregular nebulae) and sparkle with hot giant and supergiant Population I stars. Much nearer than M31 are the two Magellanic Clouds. About 200,000 light years away, they are between a third and a half the size of the Galaxy. Fairly irregular in shape, they have no definite spiral structure and consist of what appears to be a chaotic mixture of stars, gas and dust. Recent observations with radio telescopes show that they are linked by an extremely thin but common envelope of hydrogen gas. The Galaxy, the two Magellanic Clouds, M31, M33 and at least 12 other galaxies form what is known as the *Local Cluster*. All its members lie within a volume of space about 4 million light-years across. As a cluster it is quite small. The one in Virgo contains at least 1,000 galaxies while the largest, in Coma Berenices, is believed to have as many as 10,000.

Galaxies vary a great deal in shape, size and luminosity. There are three main types: elliptical, spiral and irregular. The ellipticals, with little or no structure, can be oval or round in shape. Some, like the two 'companions' of M31, are much smaller than the average spiral. Others, like M87, a giant, spherical member of the Virgo cluster, contains over 30 times the number of stars in the Galaxy. The spirals are all roughly equal in size to the Galaxy. Like the Galaxy they are highly flattened systems of stars, gas and dust, each with a nucleus in the middle. They fall into two main types: 'normal' and 'barred'. Our Galaxy, M31, M33 and M51 are all 'normal' in the sense that they have spiral arms attached to their nuclei. The 'barred' types have arms which extend from the ends of a bar that passes through the nucleus. In the irregular galaxies, stars, gas and dust are so mixed up that little or no spiral structure can be seen. Some irregulars, mere dwarfs compared with the average spiral, are just a few thousand light-years across. Their low luminosities make them difficult to detect, but they are thought to be very numerous.

Since Hubble's time the velocity-distance relationship for galaxies has been checked, rechecked and applied to objects at ever-increasing distances. As yet, the most distant galaxy photographed with the 200-

M87, a giant spherical galaxy in Virgo. Short exposure showing nuclear jet.

inch telescope is 3C-295 in Boötes. (3C stands for the Third Cambridge Catalogue of Radio Sources and 295 for the number in the catalogue). Its red-shift indicates that it is travelling away from us at about $\frac{2}{5}$ the speed of light. This velocity is, of course, quite fantastic unless we remind ourselves just how big a galaxy is. If our Galaxy travelled at this rate it would still take 250,000 years to move through a distance equal to its diameter. In comparison, the pace of a snail is like a streak of lightning. According to the velocity-distance relationship the distance of 3C-295 is 5,000 million light-years. We see it as it was 5,000 million years ago and therefore at a fairly early stage of its evolution. Similarly, an astronomer on 3C-295 would find that the Galaxy was receding at about $\frac{2}{5}$ the speed of light. All red-shifts, and therefore speeds of recession, are relative to the observer. An astronomer on *any* galaxy would find that the other galaxies were travelling away from him, and that the farther away they were the faster they would be moving. So although any one galaxy, ours included, might appear to be at rest at the centre of the universe, this is not really the case. There is no way of telling where *the* centre of the universe is—there are as many centres as there are galaxies.

Radio telescopes now play an important part in exploring the universe of galaxies. Large and highly sensitive aerials have located a great number of point-like sources of radio energy. Some of them are well-known optical objects like the Crab Nebula and the Great Nebula in Orion. A few have been pinned down to unusual stars like *UV Ceti*. These stars suddenly rise and fall in brightness. Each rise is accompanied by a burst of radio energy and is thought to be due to a great flare-like eruption. Radio sources of this kind are known as *flare stars*. But the great majority of sources appear to be outside the Galaxy. A very intense one, Cygnus A, has been indentified with two faint galaxies. They are about 700 million light-years away and look as if they are in contact. Others have been identified with similar 'contacting' pairs and also with single objects showing peculiar features such as bright jets or streamers. M87, the giant spherical galaxy in Virgo and a strong radio source, has a narrow jet or 'tail' of this kind. M82 in Ursa Major, an irregular galaxy rich in dust, shows a great central jet on photographs taken in the red light of hydrogen. Everything about these and similar objects indicates that they are in the throes of colossal explosions. But just what is exploding and how the explosions are caused is unknown.

Strangest of all are the *quasars* or *quasi-stellar radio sources*. They are easily the most intense of all radio sources. They look like stars but are far too large and luminous to be stars. Their red-shifts indicate that they are extremely distant, but the lines in their spectra are bright ones due to intensely hot gases. So they cannot be great systems of stars like the giant elliptical and spherical galaxies. Judged by its red-shift, one of them, 3C-147, may have a velocity of recession of 76,000 miles a second, nearly half the speed of light. If so, its distance is probably 6,000 million to 7,000 million light-years, greater even than that of 3C-295. Another, 3C-273, despite its great distance of over 1,500 million light-years, looks like a star of the 13th magnitude. With a luminosity 800,000 million times that of the sun it is easily the most luminous object in the heavens, yet its diameter is reckoned to be less than 10 light-years. Photographs taken with the 200-inch telescope show that a bright jet or 'tail' extends from its centre. Further, it varies in its brightness; so does another, 3C-48. All of which adds to the great mystery at present surrounding these strange objects. There is, however, little doubt that the lavish amounts of light and radio energy they send out are due to titanic explosions. If we imagine the explosion of a supernova to be like a bonfire that of a quasar is like an atom bomb.

The 200-inch telescope is thought to be capable of reaching out to at least 2,000 million galaxies. Just inside its present photographic limit are extremely faint ones like 3C-295, some 5,000 million light-years away and travelling at nearly $\frac{2}{5}$ the speed of light. What lies beyond this limit? Presumably a great number of quasars like 3C-147 and almost certainly millions more galaxies. If the velocity-distance relationship keeps its straight-line form for all distances, no matter how great, the distance is reached when a galaxy is moving at the speed of light. Its distance is then about 13,000 million light-years, but its light, dragged along behind, never reaches us. It and all its fellows are therefore forever beyond our ken. Which means, of course, that the observable universe has a definite boundary as far as we are concerned. But then an astronomer on a galaxy at the boundary of our universe would consider our Galaxy to be at the boundary of his universe. There are as many boundaries as there are galaxies.

Since the universe of galaxies is expanding in all directions, how did it begin? According to one theory, first proposed by Abbé Georges

Lemaître in 1927, all the material in the universe was once concentrated into a giant atom. About 10,000 million years ago this 'primeval atom' or 'cosmic egg' exploded and the debris gradually condensed into clusters of galaxies. So as time advances and the galaxies fly apart, space gets emptier. Very distant galaxies will appear to be more crowded together than near ones, since in looking at them we are, in effect, also looking far back in time.

Another theory, proposed fairly recently by Thomas Gold, Hermann Bondi and Fred Hoyle, holds that the number of galaxies in a very large volume of space remains the same. Although the galaxies are flying apart, the growing emptiness of space is balanced by the continuous creation of new galaxies out of nothing. On this basis the universe has always been in existence and will remain so for all time. Also, very distant galaxies will appear to be no more crowded together than nearer ones.

Which theory, the 'evolutionary' or the 'steady-state', fits the facts of observation the better? At present, opinion is divided. Martin Ryle, a radio astronomer at Cambridge, favours the evolutionary theory. His counts of over 2,000 radio sources indicate that the sources are more crowded together farther off than they are close by. Counts by radio astronomers in Australia, however, have produced different results. Nor is everyone agreed that Ryle's radio sources are all extremely distant. The quasars probably are, so they may provide the necessary evidence. But their discovery has brought along a whole new crop of problems which it may take quite a time to solve. This is how it is with science. One discovery leads to another. A problem solved immediately raises others. As a result the tide of human knowledge is always advancing into the unknown. Were it not so we should be back with the Ancient Greeks in our thinking—back to the fixed earth, the fixed circular orbits of the sun, moon and planets, and the fixed stars of a man-centred universe.

Appendix A

THE PLANETS

NAME	MEAN DISTANCE FROM SUN (MILLION MILES)	ORBITAL PERIOD (YEARS)	EQUATORIAL DIAMETER (MILES)	AVERAGE DENSITY (WATER=1)	NO. OF SATELLITES
Mercury	36·0	0·24	3,010	5·41	0
Venus	67·2	0·62	7,650	4·89	0
Earth	92·9*	1·00	7,927	5·51	1
Mars	141·5	1·88	4,220	3·95	2
Jupiter	483·3	11·86	88,760	1·33	12
Saturn	886·1	29·46	74,160	0·71	9
Uranus	1,782	84·01	29,300	1·70	5
Neptune	2,792	164·79	27,800	2·26	2
Pluto	3,664	248·43	3,700	5·5	0

* 93.50 according to radar observations of Venus (pp. 95-96).
If this value is adopted all the other mean distances will be subject to slight increases.

Appendix B

THE TWENTY BRIGHTEST STARS

CONSTELLATION	STAR	APPARENT VISUAL MAGNITUDE	LUMINOSITY (SUN=1)	DISTANCE (LIGHT YEARS)
Canis Major	Sirius*	−1·47	26	8·7
Carina	Canopus	−0·9	13,000	300
Centaurus	Rigel Kent†	−0·3	1·44	4·3
Lyra	Vega	0·0	50	26
Auriga	Capella†	0·09	150	45
Boötes	Arcturus	0·2	100	36
Orion	Rigel*	0·2	18,000	850
Canis Minor	Procyon*	0·3	5·4	11
Eridanus	Achernar	0·5	200	75
Centaurus	Hadar	0·6	3,100	300
Aquila	Altair	0·75	9	16
Orion	Betelgeuse	0·9 Var.	13,000 Var.	650
Crux	Acrux*	1·0	1,650	270
Taurus	Aldebaran	1·1	90	65
Geminus	Pollux	1·2	28	35
Virgo	Spica	1·2	1,500	220
Scorpius	Antares*	1·2 Var.	3,400 Var.	400
Piscis Australis	Fomalhaut	1·3	13·5	23
Cygnus	Deneb	1·3	50,000	1,500
Leo	Regulus†	1·3	70	85

* Double Star   † Triple Star

Glossary

OF ASTRONOMICAL TERMS

Aberration   (1) An optical defect in an optical system such as a telescope, microscope or camera.
(2) The apparent displacement in the position of a star due to the earth's velocity in its orbit and the finite velocity of light.

Absorption   The partial or complete cutting off of light by material placed in its path, e.g., the atmospheres of planets and stars, interstellar dust and gas.

Achromatic   A term applied to the object-glass and eyepiece of a telescope when they have been corrected for false colour or chromatic aberration.

Aerolite   A meteorite made up largely of stone, or silicate minerals.

Angström unit   A unit of wavelength in the measurement of light. It is one ten thousand-millionth part of a metre.

Aperture   The effective opening of an optical system. In the case of a telescope it is usually the diameter of the object-glass.

Apex, solar   The point on the celestial sphere towards which the sun is moving.

Aphelion   The point in an orbit of a planet or comet which is farthest from the sun.

Apogee   The point in the orbit of a body travelling around the earth which is farthest from the centre of the earth.

Astronomical unit   The mean or average distance of the earth from the sun. It is roughly 93 million miles.

Binary   A system of two stars each of which revolves about the common centre-of-mass of the system.

Celestial equator   The projection of the earth's equator on the celestial sphere.

Celestial poles   The two points where the earth's axis, continued into space, meets the celestial sphere.

Celestial sphere   An imaginary sphere on which the stars appear to be

GLOSSARY

Centrifugal effect   fixed. It is centred on the observer and has an infinitely large radius.

Centrifugal effect   The tendency of a revolving body to fly away from the centre of its path or orbit.

Cepheid   A star whose light varies in a way similar to that of the star *Delta Cephei*. The period, or time for one complete light change, can be anything from about 1 to 45 days. The period is proportional to the luminosity. So by measuring the period of a cepheid astronomers can determine its luminosity and, thence its distance.

Chromosphere   The lower layer of the sun's atmosphere. It consists largely of hydrogen and is red in colour.

Collimator   An optical device for producing a beam of parallel rays of light.

Conjunction   A near apparent approach of two or more celestial objects. Inferior conjunction occurs with Mercury and Venus when they lie between the earth and the sun. Superior conjunction occurs when they lie on the far side of the sun.

Coriolis force   The name given to deviating effect produced on a body moving on the surface of a rotating globe.

Corona   The outer part of the sun's atmosphere. During a total eclipse of the sun it appears as a pearly-white halo of light.

Coronagraph   A telescope designed to photograph the sun's corona and prominences in broad daylight.

Cosmic rays   Fast-moving electrified particles which come from outer space.

Day   The period of the earth's rotation, not be confused with daytime, or the interval between sunrise and sunset. Measured relative to the stars the length of the day is about 23 hr. 56 min. 4 sec.

Density   The mass of a body divided by its volume.

Dispersion   The separation of white light into its constituent colours when it is bent or refracted by a single lens or glass prism.

Doppler-Fizeau effect   A shift in the spectrum lines of a body moving towards or away from an observer. The shift is measured relative to the corresponding lines in the spectrum of a laboratory source.

GLOSSARY · 149

Dwarf stars — Stars of the Main Sequence.

Eclipse — The passage of one celestial body through the shadow of another. Thus an eclipse of the sun occurs when the earth passes through the moon's shadow and one of the moon occurs when the moon passes through the earth's shadow.

Eclipsing binary — A system of two stars whose components periodically occult or cut off part or all of the light of each other.

Electromagnetic radiation — Radiation of all known wavelengths, from extremely short gamma rays through x-rays, ultra-violet, visible light and infra-red to long radio waves.

Electron — A particle of unit negative charge.

Ellipse — A geometrical curve drawn round two fixed points called foci. The sum of the distances between the foci and any point on the curve is constant and equal to the greatest diameter of the ellipse.

Equinox — The time when the sun crosses the celestial equator.

Eyepiece — The lens or combination of lenses placed at the eye end of a telescope. Its purpose is to magnify the image formed by the object glass.

Faculae — Bright spots and streaks on the surface of the sun. They are probably mountains of hot gas.

Flare, solar — A brief but brilliant eruption on the sun with associated emissions of radio waves, ultra-violet radiation and streams of electrified particles.

Fluorescence — The absorption of radiation of short wavelengths by a substance and its emission at longer wavelengths.

Focal length — The distance, measured along the optical axis, between the surface of a lens or mirror and its focus.

Focus — The point on the optical axis of a lens or mirror at which parallel rays of light converge after refraction or reflection.

Galaxy, the — The Milky Way System, a system of about 100 thousand million stars of which the sun is a member. The Galaxy is one among several thousand million similar systems.

Gamma rays — Electromagnetic radiation of extremely short wavelength.

Geodesy — The study of the shape and size of the earth.

Geoid — The shape of the earth based on the average level of the seas and oceans.

GLOSSARY

Giant stars — Stars more luminous and therefore larger than main sequence stars of the same spectral type.

Gravitation — The tendency of all bodies in the physical universe to attract one another with a force directly proportional to the product of their masses and inversely proportional to the square of their distance apart.

Gravity — The pull of the earth on objects on its surface.

Gravity, centre of — The point where the entire mass of a body or system of bodies may be considered to be concentrated and to act.

Gyroscope — A heavy wheel mounted so that it can spin rapidly about an axis which can be set to point in any direction. It is used as a steering aid and stabilizer in ships and aircraft.

Infra-red — A band of electromagnetic radiation beyond the red end of the spectrum but shorter in wavelength than radio waves. It has important heating properties.

Interferometer — A device used in optical and radio astronomy for measuring extremely small angles such as the apparent diameters of stars.

Ionisation — The process in which an atom loses one or more of its electrons. Since an electron is a unit charge of negative electricity the atom becomes positively charged, or ionised.

Ionosphere — A region of the earth's upper atmosphere in which most of the atoms are ionised.

Laser — An optical device for producing an extremely intense parallel beam of light.

Latitude — The angular distance of a place north or south of the equator.

Librations — Apparent swinging movements or oscillations of the moon on its axis. They are due to the inclination of the moon's orbit and to its varying velocity.

Light-year — The distance that light travels in a year at a speed of 186,262 miles a second. It is roughly 6 million million miles.

Longitude — The angular distance east or west of the meridian of Greenwich.

Luminosity — The actual brightness or light output of a star as opposed to its apparent brightness.

GLOSSARY · 151

Lunar probe — A small space-vehicle equipped with measuring instruments and/or cameras for finding out more about the moon.

Magnitude, absolute — The magnitude a star would have if it were at a standard distance of 10 parsecs or 32·6 light-years.

Magnitude, apparent — The measure of the apparent brightness of a star or planet. A change of 5 magnitudes corresponds to a hundred-fold change in brightness.

Magnetosphere — A region of particle radiation surrounding the earth. The radiation consists mainly of electrons and protons temporarily trapped in the earth's magnetic field.

Main sequence — If the stars near the sun are arranged in order of increasing luminosity most of them form a sequence of increasing surface temperature and size. The sun lies on this sequence and is therefore called a main sequence star.

Mass — The quantity of matter in a body.

Meridian — Any great circle passing through the poles of the earth.

Meteor — A 'shooting star', or small particle which coming at great speed from outer space, is heated and made to shine by friction with the air in the earth's upper atmosphere.

Meteorite — A solid body of stone, iron, or a mixture of both, which may or may not fall from outer space to the earth's surface.

Micrometer — An instrument used for measuring small angular distances, e.g. the apparent diameters of planets and the angular separations of the components of double stars.

Microwaves — The part of the electromagnetic spectrum between the long infra-red waves and short radio waves.

Monochromator — An optical filter designed to transmit light of almost a single wavelength.

Nutation — A regular to and fro movement or nodding of the earth's axis additional to the slow uniform motion of precession.

Object-glass — The front lens of a telescope.

Oblate spheroid — The figure obtained when an ellipse is rotated on its shorter or minor axis.

Occultation — The disappearance of a star or other body when it passes behind a nearer body of larger apparent size.

152 · GLOSSARY

	A total eclipse of the sun is really an occultation of the sun by the moon.
Opposition	A planet, comet or the moon is said to be in opposition when its direction in the sky is opposite to that of the sun.
Parallax	The apparent change in the position of a distant object caused by a change in the observer's position. *Annual parallax* is the apparent change in star positions brought about by the earth's revolution around the sun. *Horizontal parallax* is the angle subtended at an object in the solar system (e.g. sun, moon, a planet or comet) by the equatorial radius of the earth.
Parsec	A distance corresponding to an annual parallax of $1''\cdot 0$ of arc. It is equivalent to 3·26 light-years.
Penumbra	Part shadow, or the outer region of the shadow of the moon or of a planet where some light is still received from the sun.
Perigee	The point on the orbit of the moon, or of an artificial satellite, which is nearest to the centre of the earth.
Perihelion	The point on the orbit of an object in the solar system which is nearest to the sun.
Perturbations	Disturbances in the motion and orbit of a body in the solar system produced by the gravitational attractions of other bodies.
Photoelectric cell	A device which can turn the energy of light into electrical energy.
Photometry	The comparison of the brightness of a light source, such as a star, with that of a standard source.
Precession	The slow uniform wobble of the earth's axis. One wobble is made in about 26,000 years.
Prominences	Clouds, jets and streamers of hot gas in the lower atmosphere of the sun.
Proper motion	The gradual change in the position of a star due to its actual motion in space.
Proton	The nucleus of the hydrogen atom.
Quadrant	An instrument used in early times for measuring the angular distance of a celestial body from the zenith.
Radar	A radio method of finding the distances of the moon and planets. It depends on the time taken by a radio signal to reach the body in question and be reflected back again.

GLOSSARY · 153

	The time in seconds, multiplied by the velocity of light (186,262 miles per second), is equal to twice the distance of the object in miles.
Radiant	The point on the celestial sphere from which the tracks of meteors in a shower of meteors appear to diverge.
Radiation	Electromagnetic energy in the form of rays or waves, e.g., x-rays, light, heat, radio waves. The term is also applied to streams of tiny electrified particles such as electrons and protons. Radiation of this second kind is called 'particle radiation'.
Radiometer	An instrument for detecting and measuring radiation.
Red-shift	A shift towards the red of the lines in the spectrum of a source which is travelling away from the observer.
Reflector	A reflecting telescope in which the main or primary image is formed by a spherical or paraboloidal mirror.
Refraction	The bending of a ray of light when it passes from one transparent medium to another.
Refractor	A refracting telescope in which the main or primary image is formed by a lens called the object-glass.
Resolving power	The ability of an optical instrument to show fine detail or to separate the images of two very close objects.
Retrograde motion	Motion in a sense or direction opposite to the main or general direction of motion.
Siderite	A meteorite composed mainly of nickel and iron.
Siderolite	A meteorite composed of about equal amounts of iron and stone.
Spectrogram	A photograph of a spectrum.
Spectrograph	An instrument for obtaining photographs of spectra.
Spectroheliograph	An instrument with which the sun can be photographed in the light of a single spectral line.
Spectroscope	An instrument for seeing spectra.
Spectroscopic binary	A star known to be a binary by the regular doubling of the lines in its spectrum. The doubling, a Doppler-Fizeau effect, is produced by the regular to-and-fro motions of the two stars as they travel round the centre of gravity of the system.
Spectrum	The rainbow-coloured band of light formed when white light is dispersed by a prism.

Supergiant stars	Stars of immense size and high luminosity compared with the sun and other main sequence stars.
Supernova	An exceptionally bright nova or 'temporary star'.
Telluric lines	Absorption lines in spectrum of the sun formed by oxygen, carbon dioxide and other gases in the earth's atmosphere.
Terminator	The line between the bright and dark sides of the moon or a planet.
Theodolite	A surveying instrument used to measure horizontal and vertical angles.
Thermocouple	An electrical device for measuring temperature, used to measure the heat received from a distant object such as a planet or star.
Transit	The passage of a smaller body across the disc of a larger one, e.g., of Mercury or Venus across the sun. The term is also applied to the passage of a body across some fixed reference line like the celestial meridian.
Triangulation	A trigonometric method of surveying based on a series of triangles.
Ultra-violet	That part of the spectrum beyond the violet which joins up with the x-rays. It has important chemical effects.
Umbra	Deep shadow, or the inner region of the shadow of the moon or of a planet where no light at all is received from the sun.
Van Allen belts	Two belts or zones in the earth's magnetosphere in which the energy of the trapped particles is particularly high.
Velocity	The speed of a body in a particular direction. Radial Velocity is the speed directly towards or away from the observer.
Wavelength	The length of a wave of radiation measured from one crest to the next. Also, the distance radiation travels during one complete vibration of the source.
X-rays	Radiation of very short wavelength beyond the ultra-violet.
Zeeman effect	The splitting of each spectral line into two or more lines when the source is in a strong magnetic field.
Zenith	The point on the celestial sphere directly overhead.
Zodiac	The band or belt on the celestial sphere in which move the sun, moon and planets.

Bibliography

Alter, D.	*Pictorial Guide to the Moon*, Thomas Crowell Co., 1963.
Baker, R. H.	*Astronomy*, van Nostrand, 1959.
Barlow, C. W. C. and Bryan, G. H.	*Elementary Mathematical Astronomy*, University Tutorial Press, 1956.
Boyd, R. L. F.	*Space Research by Rocket and Satellite*, Arrow Science, 1960.
Fielder, G.	*Structure of the Moon's Surface*, Pergamon Press, 1961.
Graham Smith, F.	*Radio Astronomy*, Penguin, 1960.
Inglis, S. J.	*Planets, Stars and Galaxies*, John Wiley, 1961.
King, H. C.	*Exploration of the Universe*, Secker and Warburg, 1964.
	The History of the Telescope, Charles Griffin, 1955.
King-Hele, D.	*Satellites and Scientific Research*, Routledge and Kegan Paul, 1960.
Menzel, D. H.	*Our Sun*, Harvard Books, Blakiston, 1959.
Moore, P.	*Survey of the Moon*, Eyre and Spottiswoode, 1963.
	The Planets, Eyre and Spottiswoode, 1962.
Muirden, J.	*The Sun's Family*, Weidenfeld and Nicolson, 1962.
Nininger, H. H.	*Out of the Sky: An Introduction to Meteoritics*, Dover, 1959.
Norton, A. P.	*Norton's Star Atlas and Telescopic Handbook*, Gall and Inglis, 1964.
Pannekoek, A.	*A History of Astronomy*, Allen and Unwin, 1960.
Roth, G. D.	*The System of Minor Planets*, Faber and Faber 1962.
Struve, O.	*Elementary Astronomy*, Oxford University Press. 1959.

Index

Achromatic lens, 41, 45
Adams, John Couch (1819-92), English astronomer, 84
Adams, Walter Sydney (1876-1956), American astronomer, 134
Adonis, 83
Aerolites, 94
Aldebaran, 108, 128, 135
Algol, 136
Al-Khwarizmi, 12, 13
Allegheny Observatory, 78
Alpha Crucis, 136
Altair, 113
America, South, 12
Ammonia, 75, 76, 77, 79, 84, 90
Andromeda, 91, 121, 130, 137
Ångström, 48
Antares, 133, 134, 135
Antoniadi, E. M., astronomer, 66
Aphelion, 30, 31, 66, 71, 83, 88, 96
Apogee, 53, 55
Apollo, 83
Arcturus, 108, 110, 135
Arecibo radar-radio telescope 66, 69
Ariel, 80
Aristarchus of Samos, (fl. 280-264 B.C.), Greek astronomer, 19, 26, 52; crater, 63
Aristotle (384-322 B.C.), Greek philosopher, 10, 17, 18, 19, 20, 21, 26, 27, 29, 37, 38, 39, 87, 108
Asteroid, 81
Astraea, 82
Atom(s), 120, 122, 133, 137, 143; bomb, 105, 143; ionised, 103; nuclei of, 105
Atomic Physics, 137; power-station, 137
Aurorae, 105, 106, 107

Baade, Walter, astronomer, 83, 137, 140
Barnard, Edward Emerson (1857-1923), American astronomer, 76

Berlin Observatory, 84
Bessel, Friedrich Wilhelm (1784-1846), German mathematician and astronomer, 111
Beta Persei, 136
Betelgeuse, 133, 134, 135
Binary system, 109, 135
Bond, George Philips (1825-65), American astronomer, 77
Bondi, Hermann, astronomer, 144
Boötes, 142
Bradley, James (1693-1762), English astronomer, 34, 35
Brahe, Tycho (1546-1601), Danish astronomer, 29, 31, 32, 65, 87
Bunsen, Robert Wilhelm (1811-99), German chemist and physicist, 43, 44; burner, 103

Calcium, 102, 126, 127, 133
Callisto, 76
Canes Venatici, 132
Capella, 113, 136
Carbon, 89, 133, 137; dioxide, 50, 51, 68, 69, 72, 90
Carrington, Richard Christopher (1826-75), English astronomer, 98, 105
Cassini, Giovanni Domenico (1625-1712), Italian astronomer, 39, 77
Centaur (Centaurus), 111, 112
Cepheids, 113, 114, 115, 116, 118, 131, 138, 140
Ceres, 81, 82
Chromosphere, 101, 102, 103
Columbus, Christopher (1451-1506), reputed discoverer of the New World, 12
Coma, 90; Berenices, 127, 140
Comet(s), 87-94, 121; Biela, 91; Encke, 92; Giacobini-Zinner, 92; Halley, 87, 88, 90, 92; Morehouse, 89; Mrkos, 89, 92; Swift-Tuttle, 92; Tempel-Tuttle, 91; Wilson-Harrington, 92
Compass needle(s), 105
Continuous spectra, 44
Copernicus, Nicolaus (1473-1543), Polish astronomer, 19, 20, 26, 28, 29, 31, 32, 37, 38, 39, 76
Coriolis force, 23
Corona, 101, 102, 103, 107; inner, 102, 103
Coronagraph, 102
Cosmic dust, 104; rays, 106
Crab, 123, 127, 142
Cyanogen, 89
Cygnus, 111, 112, 142

Dawes, W. R., English astronomer, 77
Deimos, 74
Della Porta, Giovanni Battista (c. 1538-1615), Italian natural philosopher, 36
Delta Cephei, 114, 140
Delta Orionis, 126
Deneb, 113
Doppler-Fizeau effect, 47, 66, 69, 78, 84, 98, 108, 115, 120, 123, 132
Draco, 44, 92
Draper, Henry (1837-82), American scientist, 44
Dust, 121-32
Dwarfs, See Sequence

Earth, axis, 26, 35, 95; flat, 9; mantle, 16; radius, 14; rotation of, 17-25
Earthquake waves, 25
East, Middle, 9; Near, 9
Eclipse, 33; lunar, 10, 31; solar, 24, 31, 101, 103; total, 102
Eddington, Sir Arthur Stanley (1882-1944), English astronomer, 136
Egypt, 9, 10
Electromagnet(s), 101
Electron(s), 103, 105, 107, 113, 120, 122, 137

156

INDEX

Equator, 10, 14, 20, 22; celestial, 35
Equinoxes, Precession of, 35
Eratosthenes of Alexandria (*c*. 276-194 B.C.), Greek writer and mathematician, 10, 11, 12, 13
Eros, 83
Eta Aquarids, 92
Europa, 76

Faculae, 97, 105
Fluorescence, 122, 124
Foucault, Jean Bernard Léon (1819-65), French physicist, 23, 24
Fraunhofer, Joseph von (1787-1826), German physicist, 43, 111

Galactic centre, 117, 129; clusters, 127, 128; plane, 118, 126,, 130, 131
Galaxies, elliptical, 137, 138, 140, 143; irregular, 140, 141;. spherical, 143; spiral, 137, 138, 140
Galaxy, 47, 111, 112, 116, 117, 118, 119, 120, 126, 127, 128, 129, 130, 131, 132, 133-44; Great, 137
Galilei, Galileo (1546-1642), Italian astronomer, 32, 36, 37, 38, 39, 55, 65, 76, 98, 108, 121
Galle, Johann Gottfried (1812-1910), German astronomer, 84
Ganymede, 76
Gas(es), 44, 45, 46, 54, 75, 84, 90, 98, 120, 121-32, 140, 143; cosmic, 138; sodium, 43
Gascoigne, William, 41, 42
Gassendi, or Gassend, Pierre (1592-1655), French philosopher and mathematician, 65
Geodesy, 16
Giants, *See* Sequence
Globular (star) clusters, 116, 118, 120, 121, 128, 130, 138, 140
Gold, Thomas (1920-   ), American astronomer, 144
Gravitation, 22; Inverse Square law of, 31, 32, 35, 87, 135; Newton's theory of, 14, 96

Gravity, 54, 109; force, 14, 16
Great Nebula, *See* Nebula(e)
Gyroscope, 24, 35

Hadley, George, 22, 23
Hale, George Ellery (1868-1938), American astronomer, 99, 100, 101
Hale telescope, 41, 86
Hall, Asaph (1829-1907), American astronomer, 74
Hall,. Chester Moor, (1703-71), English inventor, 41
Halley, Edmond (1656-1742), English astronomer and mathematician, 67, 87, 108, 121
Harrison, John (1693-1776), English horologist, 15
Hartmann, G., German astronomer, 126
Harvard College Observatory, 133
Heliometer, 111
Helium, 43, 75, 102, 127, 133, 137, 138; ionised, 122
Hemisphere, northern, 22, 52, 100; southern, 16, 23, 52, 100, 128
Henderson, Thomas (1798-1844), Scottish astronomer, 111, 112
Heracleides Ponticus, (*fl*. 4th. cent. B.C.), Greek philosopher, 19
Hercules, 109, 121, 129
Hermes, 83
Herschel, Sir Frederick William (1738-1822), English astronomer, 41, 79, 80, 109, 110, 111, 121, 122, 131
Hertzsprung, Enjar, Danish astronomer, 134
Hevelius, Johannes (1611-87), German astronomer, 39, 65
Hipparchus (*fl*. 160-125 B.C.), Greek astronomer, 31, 34, 52, 112
Horizontal parallax, 31, 32
Horrocks, Jeremiah (1619-41), English astronomer, 67
Hoyle, Fred (1915-   ), British astronomer, 144
Hubble, Edwin Powell (1889-1953), American astronomer, 130, 131, 132, 140, 141
Huggins, Sir William (1824-1910), British astronomer, 44, 45, 121, 122
Huygens, Christian (1629-93), Dutch physicist, 37, 39, 78, 121
Hyades, 127, 128
Hydrogen, 75, 76, 77, 89, 99, 102, 103, 127, 133, 137, 138, 140, 142; atom, 120; gas, 90, 107, 120; ionised 122, 124

Icarus, 83
Inferior conjunction, 38
Inverse Square law of Gravitation, *See* Gravitation
Io, 76
Ionosphere, 49, 69, 105
Iron, 133

Jeffreys, Harold (1891-   ), English geophysicist, 75
Jodrell Bank, 95
Jones, Sir Harold Spencer, Astronomer Royal, 83
Juno, 81, 82
Jupiter, 9, 18, 27, 28, 32, 33, 37, 39, 47, 51, 54, 65, 74, 75, 76, 77, 78, 79, 81, 82, 83, 87, 90, 91, 96; Great Red Spot, 74, 75, 78; moons of, 32; satellites of, 39, 76

Keeler, James Edward (1857-1900), American astronomer, 78
Kepler, Johannes (1571-1630), German astronomer, 29, 30, 31, 55, 65, 66, 81, 83, 92, 135
Kirchhoff, Gustav Robert (1824-87), German physicist, 43, 44
Kirkwood, Daniel (1814-95), American astronomer, 92
Kozirev, Russian astronomer, 62, 63
Kuiper, Gerard P., 80, 84, 86

Laser, 95
Lassell, William (1799-1880), British astronomer, 80, 84
Leavitt, Henrietta S., astronomer, 114
Lemaître, Abbé Georges Édouard (1894-   ), Belgian astrophysicist, 143

INDEX

Leo, 90
Leonids, 90, 91
Leverrier, Urbain Jean Joseph (1811-77), French astronomer, 84, 91
Lexell, mathematician, 79
Libration in latitude, 55
Libration in longitude, 55
Lick Observatory, 118
Light-year, 112, 113, 127, 128, 130, 134, 140, 141, 142
Lippershey, Hans, Dutch inventor, 36
Local cluster, 140
Lowell Observatory, 72, 86
Lowell, Percival (1855-1916), American astronomer, 73, 85
Lunar day, 55
Lunar probe, 63, 64
Lunik I, II and *III*, 63
Lyot, Bernard Ferdinand, French astronomer, 102
Lyrae, 116, 118, 122, 138, 140
Lyttleton, Raymond A., English astronomer, 90

McDonald Observatory, 80
Mädler, Johann Heinrich von (1794-1874), astronomer, 56, 58
Magellan, Ferdinand (*c*.1480-1521), Portuguese navigator, 13
Magellanic Clouds, 137, 140
Magnesium, 133
Magnetic equator, 107; field, 100, 101, 105, 106; force, 102; polarities, 100; poles, 103, 106; storms, 105
Magnetosphere, 107
Magnitude, Absolute, 113, 114, 115, 116, 134; Apparent, 112, 132; 134; Photoelectric, 113; Photographic, 113
Mare(s), Nubium, 64
Maria, 54, 58, 59, 61, 63, 66
Mariner 2, 69, 70
Mariner 4, 74
Marius, Simon (1570-1624), German astronomer, 121
Mars, 9, 18, 27, 28, 29, 31, 47, 51, 65, 70, 71, 72, 73, 74, 81, 82, 83, 96, 135; canals, 73; polar caps, 71
Maupertuis, Pierre Louis Moreau de (1698-1759), French mathematician and astronomer, 15
Maxwell, James Clerk (1831-79), Scottish physicist, 77
Mercury, 9, 18, 27, 28, 38, 65, 66, 67, 68, 69, 70, 83, 86, 96
Meridian, 12, 13, 24, 52
Messier, Charles, French astronomer, 121, 130
Messier numbers, 121, 122, 123, 124, 126, 128, 130, 131, 132, 137, 140, 141, 142
Meteorites, 58
Meteors, 87-94
Methane, 75, 77, 79, 84, 86, 90
Micrometeorites, 60, 93, 94
Micrometer, 76, 81, 82; eyepiece, 42
Microwaves, 61, 69
Milky Way, 37, 51, 108, 111, 112, 116, 118, 126, 127, 132
Miranda, 80
Monochromator(s), 102, 105
Moon, 52-64, 95, 101; eclipse 10, 52, 60; gravitational pull, 25; worship, 9
Mount Palomar Observatory, 41, 67, 83, 140
Mount Wilson Observatory, 41, 83, 116, 130, 134, 137

Nebula(e), 47, 122, 126, 129, 130, 131, 138; dark, 118, 121, 126; irregular, 124, 126, 140; Great, 121, 124, 130, 142; planetary, 122, 123, 124, 126, 138; reflection, 126, 130
Neon, ionised, 122
Neptune, 84, 85, 86, 88
Nereid, 84, 85
Newton, Sir Isaac (1642-1727), English philosopher, 13, 14, 15, 22, 31, 35, 39, 40, 43, 87, 135; Theory of Gravitation, *See* Gravitation
Nicholson, Seth Barnes (1891-), American astronomer, 67
Nitrogen, 49, 72, 89, 107; ionised, 122
North Pole, 20
North Pole Star, *See* Polaris
Northern Lights, *See* Aurorae
Nova(e), 21, 130, 137, 138
Nova Aquila 1908, 22
Nova Herculis 1934, 22
Nutation, 35

Oberon, 80
Object-glass(es), 39, 41, 45, 102
Olbers, Heinrich Wilhelm Matthäus (1758-1840), German physician and astronomer, 81
Omega Centauri, 116, 128
Oort, Jan H. (1900-), Dutch astronomer, 117, 118
Optical interferometer, 134
Orion, 19, 121, 124, 126, 142
Orionids, 92
Oxygen, 46, 47, 49, 72, 89, 107, 122, 127; ionised, 124
Ozone, 48

Pallas, 81, 82
Parsec, 112, 113
Particles, 105, 107; electrified, 106
Penumbra, 99
Perigee, 53, 55
Perihelion, 30, 31, 66, 86, 88, 96
Perseids, 92
Perseus, 127
Pettit, Edison (1890-), American astronomer, 67
Philippine Islands, 13
Phobos, 74
Phoebe, 78
Photosphere, 96, 97, 98, 99, 101, 102, 103
Piazzi, Giuseppe (1746-1826), Italian astronomer, 81
Picard, Jean (1620-82), French astronomer, 13
Pic-du-Midi Observatory, 102
Pickering, William Henry (1858-1938), American astronomer, 78, 85
Planetoïd(s), 81, 82, 83
Planets, 65-78
Pleiades, 126, 127, 128
Plough, 19, 128
Pluto, 70, 86, 92
Polaris, 10, 12, 20, 34, 35
Population I, 137, 138, 140
Population II, 138, 140
Praesepe, 127
Prominence(s), 102
Proper motion, 108
Proton(s), 61, 103, 106, 107, 120
Proxima Centauri, 120
Ptolemy (*c*. A.D. 90-168), Egyptian astronomer and

INDEX · 159

geographer, 19, 20, 26, 30, 31, 52, 108
Pythagoras (c. 582-507 B.C.), Greek philosopher and mathematician, 10, 19

Quasar(s), See Radio

Radar, 95
Radial velocity, 47, 109, 120, 130, 132
Radiation, ultra-violet, 105, 122; x-ray, 105
Radio, astronomy, 51; "Noise", 103; quasi-stellar radio sources, 143, 144; reception, 105, 106; sun, 103
Radiometers, 60, 67, 69, 77
Radius, equatorial, 16; polar 16; polar flattening, 16
Ramsey, William, chemist 75, 76
Ranger 7, 8 and *9*, 64
Rays, See Waves
Reflector, 40, 121
Refractor, 39, 41
Reinmuth, K. German astronomer, 83
Richer, Jean, French astronomer, 14, 22
Rigel, 110, 133, 136
Ritchey, George Willis, American astronomer, 130
Römer, Olaf (1644-1710), Danish astronomer, 33
Rosse, William Parsons, third earl of (1800-67), Irish astronomer, 121
Royal Observatory, 111
Russell, Henry Norris (1877-1957), American astronomer, 134
Ryle, Martin (1918-   ), British astronomer, 144

Sabine, Sir Edward (1788-1883), British scientist, 44 105
Sacramento Peak Observatory, 102
Sagittarius, 51, 116, 118
Satellites, 32, 33; artificial, 16, 32, 47, 51, 106
Saturn, 9, 18, 27, 28, 32, 37, 47, 54, 76, 77, 78, 79, 83, 91;

moons of, 39; rings of, 39; satellites, 78
Schiaparelli, Giovanni Virginio (1835-1910), Italian astronomer, 66, 91
Schwabe, Heinrich Samuel (1789-1875), German astronomer, 99
Sequence, dwarf, 134, 137, 138; giants, 135; main, 134, 135; red giants, 137; supergiants, 135, 137; white dwarfs, 135, 137, 138
Shapley, Harlow (1885-   ), American astrophysicist, 116, 117, 118, 129
Siderites, 94
Siderolites, 94
Sinton, W., American astronomer, 72
Sirius, 43, 65, 108, 110, 111, 112, 113, 128, 135
Small Magellanic Cloud, 114
Snell, Willebrord (1591-1626), Dutch astronomer and mathematician, 13
Sodium, 127
Solar, apex, 109, 117; flare(s), 103, 105; observatory, 51; spectrum, 44, 46, 51, 101; system, 27, 28, 31, 32, 79-86, 87, 89, 92, 93, 104, 109, 119, 124, 126; wind, 90, 103, 107
Solstices, 10
Southern Lights, 105
Space probe(s), 106
Spectrogram, 44, 45, 47, 62, 72, 75, 77, 79, 89, 99, 123, 124, 126; ultra-violet, 51
Spectrograph, 43-51, 68, 72, 98, 99, 102, 133
Spectral type, 133, 134, 135
Spectroheliograph, 99, 102
Spectroscope, 45, 121, 122
Spectroscopic binaries, 136; parallax, 134
Spectrum, 39, 43, 45, 47, 48, 68, 84, 89, 98, 99, 103, 108, 121, 124, 130, 132, 133, 134, 136; electromagnetic, 48; lines, 123; solar, See Solar
Spica, 103, 133, 136
Spicules, See Spica
Stars, 18, 44, 47, 50, 52, 79, 87, 108-20, 121-32, 133-44; clusters, 32, 121, 127, 129, 138; double, 109, 110; flare, 142; red, 51

Stellar day, 24; parallax, 34, 109
Stratoscope II, 51
Strong, John S., American astronomer, 68
Struve, Friedrich Georg Wilhelm (1793-1864), German astronomer, 111
Sun, 95-107; eclipse, 24; poles of, 98; worship, 9
Sunspot(s), 38, 39, 97, 98, 99, 100, 101, 103, 104, 105; cycle, 99, 100, 101, 103, 105
Supergiants, See Sequence
Superior conjunction, 38
Supernovae, 22, 29, 124, 138, 143
Surveyor, 64

Taurus, 123
Telescope, 24, 32, 33, 34, 36-42, 43, 45, 50, 54, 55, 63, 66, 67, 75, 76, 82, 84, 85, 90, 96, 102, 111, 116, 121, 126, 127, 132, 134, 135, 140, 141, 143; radio, 51, 61, 66, 69, 95, 103, 119, 120, 140, 142; reflecting, 51, 79, 110; refracting, 56
Telluric lines, 46, 47, 68
Terminator, 37, 55, 59
Thermonuclear reaction, 137
Theta Orionis, 121
Titan, 39, 78
Titania, 80
Titanium, 133; oxide, 133
Tombaugh, Clyde William (1906-   ), American astronomer, 86
Toucan(i), 128
Transit instrument, 24
Transits, 65, 67
Trapezium, 126
Triangulation, 13, 16
Triangulum, 131
Triton, 84
Trumpler, Robert Julius (1886-1956) American astronomer, 118, 132
Tycho, 55, 56

Umbra, 99
Umbriel, 80
Uranus, 79, 83, 84, 85, 86, 135
Ursa Major, 132, 142
UV Ceti, 142

Vacuum thermocouples, 60, 72
Van Allen belts, 107
Vega, 35, 44, 111, 112
Venus, 9, 18, 21, 27, 28, 37, 38, 43, 47, 54, 65, 67, 68, 69, 70, 83, 87, 95, 96, 123, 124; transits, 67
Vesta, 81, 82
Virgo, 132, 140, 141, 142
Visual magnitude, 113; photometer, 113

Waves, gamma, 48, 51; radio, 48, 50, 51, 95, 120; ultra-violet, 48, 50, 51
Whipple, Fred (1906-   ), American astronomer, 90, 93
Wildt, Rupert, German astronomer, 75
Witt, German astronomer, 83
Wolf, Maximilian Franz Joseph Cornelius (1863-1932), German astronomer, 82

Wollaston, William Hyde (1766-1828), English chemist and natural philosopher, 43

X-rays, 48, 51, 106

Yerkes Observatory, 41

Zeeman effect, 99
Zodiac, 28, 86, 104
Zodiacal light, 103, 104
Zond-3, 63

Acknowledgements

THE Publishers would like to thank the following for kind permission to use photographs as noted below.

Dominion Astrophysical Observatory, Canada, *p. 129*. Lick Observatory, *pp. 56, 104, 141*. Lowell Observatory, *p. 88*. Mount Wilson and Palomar Observatories, *pp. 57, 77, 97, 100, 119, 122, 123, 125, 130, 139*. National Aeronautics and Space Administration, U.S.A., *pp. 64, 73*. Royal Observatory, Edinburgh, *p. 127*. Sacramento Peak Observatory, *p. 101*.